Teaching Your Kids New Math, 6-8

T0274264

by Kris Jamsa, PhD[2]

for dummies®
A Wiley Brand

Teaching Your Kids New Math, 6-8 For Dummies®

Published by: **John Wiley & Sons, Inc.**, 111 River Street, Hoboken, NJ 07030-5774, www.wiley.com

Copyright © 2023 by John Wiley & Sons, Inc., Hoboken, New Jersey

Media and software compilation copyright © 2023 by John Wiley & Sons, Inc. All rights reserved.

Published simultaneously in Canada

For general information on our other products and services, please contact our Customer Care Department within the U.S. at 877-762-2974, outside the U.S. at 317-572-3993, or fax 317-572-4002. For technical support, please visit https://hub.wiley.com/community/support/dummies.

Wiley publishes in a variety of print and electronic formats and by print-on-demand. Some material included with standard print versions of this book may not be included in e-books or in print-on-demand. If this book refers to media such as a CD or DVD that is not included in the version you purchased, you may download this material at http://booksupport.wiley.com. For more information about Wiley products, visit www.wiley.com.

Library of Congress Control Number: 2022951572

ISBN 978-1-119-98639-3 (pbk); ISBN 978-1-119-98642-3 (ebk); ISBN 978-1-119-98641-6 (ebk)

SKY10041566_012423

Contents at a Glance

Table of Contents

Introduction

Different people like different things. Some of us like to travel. Some like walks in the woods. And some people like pina coladas and getting caught in rain. But math? People can get emotional about math — they often either love it or hate it.

I've noticed that people who find that math is not their cup of tea never really mastered the basics. That's what this book is all about.

If you fall into the category of people who dislike "old math," you may not be excited that there is now "new math."

Relax. New math is just a result of finding better ways to solve problems. You're still going to use good ol' addition, subtraction, multiplication, and division, but you'll do it with new math. You're just going to show your kids better ways to do it, and this book will show you how.

This book provides step-by-step instructions for how to use both old and new math to solve problems that sixth- through eighth-graders must know. It also provides instructions, examples, and practice problems, and often suggests what you should say as you teach your child.

About This Book

You may have chosen this book for several reasons. If your child is struggling with math, they likely need to relearn some fundamentals. This book will help. Or, you may have a budding math genius to whom you want to teach the next set of concepts. Often, people buy this book because their kids just brought home an assignment that looks Greek to them. The good news is that this book can help you teach those concepts that the ancient Greeks passed down to us.

Regardless of your reasons, you've got the right book.

This book presents the math your child must know from sixth through eighth grade, with each chapter focusing on specific key concepts. If your child needs help with only algebra or statistics, you can turn to a specific chapter that addresses that topic. If your child is struggling at their current grade level, you can take a step back and strengthen their foundation, knowledge, and confidence from previous grades. The first section of this book provides your child with a review of the concepts they should have mastered up through fifth grade.

Within each chapter, you will find step-by-step instructions for how to teach each concept. I've also provided many example problems for you to work through with your child. Let them solve the problems right on the book's pages if you want — after all, it's your book.

I understand that it may have been a while since you solved math problems without the help of your phone's calculator app. However, I think you'll be pleased with how much math you remember! You can do this!

Foolish Assumptions

I like math. That, however, does not mean everyone does!

Don't worry if you normally turn to your phone's calculator to solve math problems. I often use mine, too!

I understand that math can be hard — Einstein, in fact, said, "Do not worry too much about your difficulties in mathematics; I can assure you that mine are still greater."

Relax. We're going to look at sixth- through eighth-grade math. You can do this. In fact, you may be surprised at how many of the math concepts in this book you can use on an everyday basis!

Don't let the terms "new math" or "Common Core math" intimidate you. This book covers standard addition, subtraction, multiplication, and division, just in new ways.

Icons Used in This Book

TIP

The Tip icon marks tips (duh!) and shortcuts that you can use to make learning new math easier, and sometimes how to know when it's time to take a break!

REMEMBER

The Remember icon points out things that you should, uh, remember! You and your child will examine a lot of topics throughout this book. I include this icon for those things you should keep in the back of your mind as you move forward.

FIND ONLINE

Throughout this book, I've included references to worksheets with additional problems your child should practice. This icon identifies those worksheets you can find on this book's companion website.

Beyond the Book

Becoming a "math whiz" requires practice. I've sprinkled many math problems your child can complete throughout this book. But because I know "practice makes perfect," I've provided many worksheets of problems on this book's companion website at www.dummies.com/go/teachingyourkidsnewmath6-8fd. As you and your child work your way through this book, you should take the time to download and print the corresponding practice worksheets.

In addition, should you need quick help on a math process and you don't have this book handy, I've created a Cheat Sheet you can download and print that will help you with many key concepts. Head to www.dummies.com and type **Teaching Your Kids New Math 6-8 For Dummies Cheat Sheet** in the search bar.

Where to Go from Here

This book's first section includes a review of the key math concepts taught through fifth grade. Often, the reason kids dislike math is because they are learning new concepts that build upon concepts that were previously taught, but which they don't fully understand. If your child is struggling with math, take time to review thoroughly this book's first section — that will be time well spent. You will not only build your child's math foundation, but also their confidence. With that, they may even start to like math!

The chapters of this book are meant to be read in order. That said, if your child drops their unfinished math homework in front of you while you are trying to drink your morning coffee, you can turn directly to the concept at hand. Later that evening, when you switch to a different beverage, you can take time to go over the concepts upon which your child's homework relies.

Most of this book's chapters introduce concepts with examples that you may find easier than your child's homework. That's intentional. I want your child to understand the concept, not struggle with larger and more difficult numbers. After your child masters a concept, you can always make the numbers larger or more complex. By then, however, your child will have the knowledge and confidence to complete such problems.

1

Securing the Foundation

Chapter **1**

Parent, Provider, and Now, Math Teacher

Many people have visions of parenthood being the joys of having a family and raising kids. There are visions of vacations on the beach, family barbecues, and happy evenings filled with board games at the kitchen table. Then, life happens. Evenings get interrupted by work calls and emails, work creeps into the weekends, and budgets that looked great on paper, never quite work out. On top of that, your pride about, and excitement for, parent-teacher conferences may get replaced by fear and nervousness at finding out that your child is learning new math and that the word "new" seems to mean only that the math is new to you!

The good news — yes, there is good news — is that you've picked up the right book! I get it! Old math may not have been one of your favorite things. In fact, you may worry that you have forgotten more old math than you remember. That's where this book fits in. It will remind you of, or possibly reteach you, the key old-math concepts. I think you will be surprised by how much you remember! This book will then teach you the new-math techniques you must know, and how to teach those concepts to your kids!

This chapter introduces you to new math — what it is and, more importantly, why you need it. It makes sense of Common Core math and how it relates to new math. Finally, this chapter helps you clarify why you have chosen to put on a math-teacher hat and gives you some pointers and hints to help you succeed.

New Math? What Was Wrong with Old Math?

When I grew up, kids played outside all day, drank water from the hose, and knew to be home before dark. I also learned "old math," which seems to have worked well for my adult needs. I know how much money I can spend, and how much change I should receive. I can recognize higher prices as well as good deals. If you had asked me a few years ago, I would have said that old math was just fine.

That said, things change over time. Many cars no longer stop for gas, but rather, to plug in. Drivers who couldn't wait to get their learner's permit when they were young, now look forward to self-driving cars. And teachers have found more effective ways to use math to solve problems.

REMEMBER

New math, simply put, provides new ways to solve problems. Carrying and borrowing to add and subtract have been replaced with number lines, and old-school multiplication has been replaced with new techniques that use boxes:

$$
\begin{array}{r}
37 \\
\times 23 \\
\end{array}
$$

	30	7
20	600	140
3	90	21

$$
\begin{array}{r}
{}^{1} \\
600 \\
+140 \\
+\ 90 \\
+\ \underline{21} \\
851 \\
\end{array}
$$

The good news is that numbers have not changed, and you still use the symbols +, −, ×, and ÷ for addition, subtraction, multiplication, and division.

All of that said, I like the new-math techniques. They are straightforward, they work, and they are fast to use. After you set aside your fears and nervousness, I think you will like them, too!

TIP

Starting out, the best tip I can offer you is to be open to new ways of learning — especially if you want to help your child master new math strategies. My goal in presenting the techniques in this book is to make it enjoyable for your child to learn new math with you. The bond you will create with your child is possibly more important than establishing their math foundation for future success.

Old Math, New Math, Common Core Math

I can remember when there was just math. It was the third "R" in Reading, wRiting, and aRithmetic. Now, when someone uses the word *math*, you must ask, "Which one: old-school math, new math, or Common Core math?"

Old-school math is the math most of us learned. Like a trusty old pickup truck, old-school math still works. As such, throughout this book, I present many old-school-math techniques that you should teach to your child.

New math includes new ways to add, subtract, and multiply numbers. Like a brand-new, shiny pickup truck, new math also works. Unfortunately, most people are introduced to new math with their child's homework assignment, which is due the following morning. Fortunately, this book will teach you what you must know to teach your child new-math techniques.

It used to be that smart people in each state would get together and establish the state's learning curriculum — the things teachers in that state would teach. The problem was that each state's curriculum was different. What a sixth-grader learned in New York might be different from what a sixth-grader learned in Arizona or Montana. Simultaneously, math scores within the United States were falling. In fact, as of 2015, math scores in the United States had fallen from first to 35th in the world!

In 2009, the National Governors Association and the Council of Chief State School Officers got together to create the Common Core State Standards Association. From that group, Common Core math was born.

If you ask a roomful of educators to comment on Common Core, you will hear a wide range of opinions. Some love it! They want to see standards across grade levels and across the country. Others hate it! They want the government to leave curriculum decisions to the individual states. This book does not debate for either side. Instead, I simply present the math skills these groups identified as important for your child to know and for teachers to teach.

REMEMBER

Common Core math encompasses the recommended math concepts that teachers are directed to teach. Because Common Core math includes many new math techniques, many people use the terms interchangeably. That said, not all Common Core math is new math — many old-school math techniques remain.

Meeting Your Child's Math Needs

I bet you're thinking, "Great! I can't wait to learn new math so I can teach it to my child!" It isn't like you didn't have anything else to do!

Given that you are reading this book, I know you are at least interested in math, or desperate because you are struggling to help your child solve their homework assignments.

The bottom line is that you can do this. You learned old-school math, and you can learn new math. In fact, you are likely to surprise yourself and amaze your friends with what you know.

TIP

If your child is struggling today, it's likely because they didn't master the skills in a previous grade level. That problem is easy to solve. This book starts with a review of key math concepts through to fifth grade. You may want to start there. Depending on your child's age and current skills, you may move through that content quickly. The successes your child will experience will give them greater confidence in their knowledge, and you may find that you're able to fill in a few key gaps. In any case, if your child is having trouble at their current grade level, you can simply turn back a few pages to a previous grade level and lay a better foundation. Remember, you paid for the entire book. Use it!

INVESTING IN YOUR CHILD'S FUTURE

Math is important. That's why schools teach math every day in every grade. Kids who do well in math tend to do well in school. Further, kids who do well in school tend to go on to college.

Research has shown that college graduates tend to earn over one million dollars more throughout their career than non-college graduates. That's a one-million-dollar return on your investment of time to help your child succeed in math.

REMEMBER

You may be worried that you are too busy to help your child with math or that you can't learn new math. Relax. You have the right book. Raising kids can be hard. The good news is that teaching math is not. You can do this!

Creating a Math Routine

Transforming your child into a math whiz takes time and effort. Following are a few tips for creating a solid math routine:

>> Plan on spending around 15 minutes a day on math with your child.

>> Try to pick a regular time each day to work with your child to establish consistency.

>> Try to pick a location that is away from other distractions, such as your television or smartphone.

REMEMBER

By establishing a routine, you will find that you can make time, and your child will have the expectation that you will be working together. Knowing that you care about their success is important to your child.

Encouraging Your Child When the Going Gets Tough

Math can be hard, and your child will make mistakes. The key is in how you respond to your child after such errors occur. A positive attitude goes a long way. Be positive about the math problems your child gets right, as well as what they can learn from the ones they get wrong.

TIP

When your child makes a mistake, and they will, make the problem the focus by saying, "This is not correct—let's look at it again," rather than saying, "You got this wrong."

Remember, your goal is to build your child's math confidence and, ideally, their enjoyment of math.

REMEMBER

Have fun! You are setting out on an adventure that will forever change your child's life.

IN THIS CHAPTER

» **Understanding mental math**

» **Handling timed addition tests**

» **Practicing addition with carrying (regrouping)**

» **Adding with a number line**

» **Cracking word problems that require addition**

Chapter **2**

Adding to What They Already Know

There's an old adage in football that winning teams perform the basics well — meaning, they know how to block and tackle. The same adage applies to math. Whether your child is solving basic math problems or is a math prodigy destined to study advanced calculus, they will always use the basics: addition, subtraction, multiplication, and division.

This chapter focuses on addition. Although some of the topics may feel like review, your goal is to improve your future math whiz's ability to solve common addition problems mentally — you know, in their heads. In addition (no pun intended), solving math problems mentally will increase their confidence. Take your time with this chapter. Many students who struggle with and thus dislike math do so because they never mastered the basics.

This chapter starts with old-school addition — meaning, your child needs to solve addition problems using carrying (sometimes called regrouping). Then, you switch to using number lines (some new math) to solve addition problems. Finally, your child gets a refresher on solving word problems that require addition. You may remember those: "Two trains leave the station" The good news is that this chapter's word problems do not have trains.

Note: Before you get started, you should buy some 3x5 index cards and a deck of playing cards.

Solving Addition Problems in Their Heads

You probably know someone who can amazingly solve math problems in their head almost as fast as you can type the problems into your phone's calculator app. In this section, your child will start to become one of those people.

To start, you create flashcards that you can use to help your child quickly add numbers through 20 in their heads.

TIP

Whoever said practice makes perfect was right. The key to success with flashcards is consistent practice. Plan to spend 10 minutes a day reviewing the flashcards with your child until they have mastered all the cards and can quickly solve the addition problems in their head.

Using your 3x5 index cards, create the following flashcards, or, if you are out shopping, you can buy flashcards:

0	0	0	0	0	0	0	0	0	0	0
+0	+1	+2	+3	+4	+5	+6	+7	+8	+9	+10
1	1	1	1	1	1	1	1	1	1	1
+0	+1	+2	+3	+4	+5	+6	+7	+8	+9	+10
2	2	2	2	2	2	2	2	2	2	2
+0	+1	+2	+3	+4	+5	+6	+7	+8	+9	+10
3	3	3	3	3	3	3	3	3	3	3
+0	+1	+2	+3	+4	+5	+6	+7	+8	+9	+10
4	4	4	4	4	4	4	4	4	4	4
+0	+1	+2	+3	+4	+5	+6	+7	+8	+9	+10
5	5	5	5	5	5	5	5	5	5	5
+0	+1	+2	+3	+4	+5	+6	+7	+8	+9	+10
6	6	6	6	6	6	6	6	6	6	6
+0	+1	+2	+3	+4	+5	+6	+7	+8	+9	+10
7	7	7	7	7	7	7	7	7	7	7
+0	+1	+2	+3	+4	+5	+6	+7	+8	+9	+10
8	8	8	8	8	8	8	8	8	8	8
+0	+1	+2	+3	+4	+5	+6	+7	+8	+9	+10
9	9	9	9	9	9	9	9	9	9	9
+0	+1	+2	+3	+4	+5	+6	+7	+8	+9	+10
10	10	10	10	10	10	10	10	10	10	10
+0	+1	+2	+3	+4	+5	+6	+7	+8	+9	+10

TIP

If you want your child to practice the flashcards on their own, write the correct answer on the back of each card so they can quickly check their result.

These flashcards all have results that are in the range 0 to 20. By helping your child master these flashcards, you are laying the foundation they will use when they perform multi-digit addition.

Mastering timed addition tests

FIND ONLINE

Your goal in practicing addition with flashcards is for your child to solve key addition problems quickly and accurately in their head. After your child has mastered the flashcards, have them practice timed worksheets like that shown in Figure 2-1, which I have provided on this book's companion website at www.dummies.com/go/teachingyourkidsnewmath6-8fd.

1 +2	2 +2	4 +3	3 +6	4 +4	5 +6	7 +3
8 +2	9 +3	4 +4	7 +6	9 +9	5 +0	7 +7
1 +8	9 +7	4 +5	7 +7	4 +8	5 +9	7 +9
1 +5	4 +9	4 +10	6 +6	5 +5	5 +8	7 +8
1 +1	2 +0	5 +3	4 +7	6 +8	1 +9	8 +3
6 +9	9 +7	6 +1	10 +6	10 +4	9 +3	10 +7
10 +10	10 +2	4 +10	1 +8	3 +3	15 +4	7 +3

FIGURE 2-1:
A timed addition worksheet.

Ideally, your child should accurately complete the worksheet in 5 minutes or less. When they do so, you can move on.

TIP

Timed tests can be stressful, so try to keep the process fun. Be positive about the problems your child gets right. If they miss several problems, that's your sign that they should spend more time with the flashcards. If you find that you're performing the tests several times with your child, you may want to create an answer sheet to simplify your review.

REMEMBER

Your child's ability to solve common addition problems in their head will build their confidence and lay the foundation for future math success.

Counting cards — so to speak

After your child can add numbers through 20 in their head, take out a deck of playing cards, removing the joker. Explain to your child that in some card games, the ace can count as 1 or 11 — for this game, it counts as 1. Explain to your child that each face card counts as 10.

Spread the cards face down on a table (or on the floor) and ask your child to turn over 2 cards. Ask them to add the cards in their head, saying the correct answer out loud. If your child is correct, they get to keep the cards; otherwise, you get to keep the cards. When all the cards have been turned over, the player with the most cards wins.

If your child easily adds the cards in the deck, you can advance the game by having your child turn over two cards and seeing which of you can say the answer faster. Whoever correctly says the answer first gets to keep the cards.

Becoming One of "Those" People Who Does Math in Their Head

It's important for their future math success that your child can quickly add numbers through 20 in their head. That said, you don't have to stop your flashcards at 20. You might, for example, add the following flashcards to your practice deck:

20 +1	20 +2	20 +3	20 +4	20 +5	20 +6	20 +7	20 +8	20 +9	20 +10

After your child has mastered these cards, you can add the following:

21 +1	21 +2	21 +3	21 +4	21 +5	21 +6	21 +7	21 +8	21 +9	21 +10

You can repeat this process of adding cards through 100.

REMEMBER

The ability to add numbers through 100 in their head will take time and practice. Do not add additional cards to your child's flashcards until they have mastered the current deck. It may take your child weeks or even months of flashcard practice to master numbers through 100. The result, however, will be worth the effort.

It's Time to Regroup

Back in the day, you learned to add by carrying, and to subtract by borrowing. These days, they call the process of carrying and borrowing "regrouping."

In this section, your child will practice addition with regrouping (you know, carrying). To start, your child will solve two-digit addition problems. To help your child write the number they are carrying, start the process by creating a box in which they can write the number As your child performs the carry operation, remind them that the digit they are carrying corresponds to a ten when they write it above the ten's column:

```
 ☐        ☐1
 3 5      3 5
+2 7     +2 7
          6 2
```

Present the following addition problems to your child, helping them as needed:

```
 ☐        ☐        ☐
 1 5      6 3      5 9
+1 8     +2 9     +2 6
```

Your child should get the following results:

```
 ☐1       ☐1       ☐1
 1 5      6 3      5 9
+1 8     +2 9     +2 6
 3 3      9 2      8 5
```

If your child does not get these results, or is not familiar with the carrying process, consider strengthening their math foundation with the book *Teaching Your Kids New Math K-5 For Dummies*, also written by yours truly and published by Wiley.

TIP Your child should correctly and quickly add the two digits in each column. If they have problems, that's a signal that you should spend more time practicing with the addition flashcards.

FIND ONLINE This book's companion website at www.dummies.com/go/teachingyourkidsnew math6-8fd contains a worksheet, the first few rows of which are shown in Figure 2-2. Your child can use it to practice two-digit addition with carrying or regrouping. Download and print the worksheet; then help your child solve the first few and ask them to complete the rest.

☐	☐	☐	☐	☐
15	23	25	38	56
+ 7	+18	+37	+13	+25

FIGURE 2-2: A worksheet for two-digit addition with carrying.

☐	☐	☐	☐	☐
37	73	42	18	37
+24	+18	+19	+13	+15

Solving three-digit addition with regrouping

In the previous section, your child practiced two-digit addition with regrouping. In this section, they perform similar operations with larger three-digit numbers. Again, as your child performs the operations, remind them that the digits they are carrying correspond to tens and hundreds.

```
 ☐☐      ①①
 322     322
+189    +189
         511
```

Present the following problems to your child, helping them as necessary:

```
 ☐☐      ☐☐      ☐☐
 194     367     819
+513    +456    +133
```

They should get the following:

☐1☐	☐11☐	☐☐1
194	367	819
+513	+456	+133
707	823	952

This book's companion website at www.dummies.com/go/teachingyourkidsnew math6-8fd contains a worksheet (the first few rows of which are shown in Figure 2-3) with which your child can practice three-digit addition with carrying. Download and print the worksheet; then help your child solve the first few, and ask them to solve the rest.

☐☐☐	☐☐☐	☐☐☐
1 3 7	3 1 7	3 6 7
+ 2 5 5	+ 2 6 9	+ 4 4 4

FIGURE 2-3:
A worksheet for practicing three-digit addition with carrying.

☐☐☐	☐☐☐	☐☐☐
1 9 2	3 8 4	1 9 3
+ 2 5 9	+ 1 6 8	+ 1 7 7

Removing the boxes

The previous addition problems use boxes to help your child add multi-digit numbers that require regrouping (I still like the term *carrying*). In this section, your child performs the same operations, but this time, without the boxes. To start, your child will begin with two-digit numbers and will then progress to numbers with three digits.

Present the following problems to your child:

38	66	18
+ 47	+ 25	+ 72

Your child should get:

☐1☐	☐1☐	☐1☐
38	66	18
+ 47	+ 25	+ 72
85	91	90

If they do not, help them add the digits in the first column, write down the number to carry (normally a 1) above the second column, and then add those digits.

This book's companion website at www.dummies.com/go/teachingyourkidsnew math6-8fd contains a worksheet (the first few rows of which are shown in Figure 2-4) that your child can use to perform two-digit addition with carrying. Download and print the worksheet; then help your child solve the first few, and ask them to complete the rest.

17	12	42	33	45	65	17
+24	+28	+39	+68	+45	+26	+23

18	19	48	73	19	58	17
+22	+33	+47	+18	+52	+13	+74

FIGURE 2-4:
A worksheet for performing two-digit addition with carrying.

19	13	14	17	34	25	17
+28	+78	+56	+73	+28	+19	+19

Do not move on to three-digit addition until your child masters the two-digit problems.

Present the following problems to your child, assisting as necessary:

347	717	814
+265	+196	+196

Your child should get the following:

11	11	1 1
347	717	814
+265	+196	+196
612	913	1010

If they do not, help them add each column of digits, writing the result and the value to carry.

This book's companion website at www.dummies.com/go/teachingyourkidsnew math6-8fd contains a worksheet (the first few rows of which are shown in Figure 2-5) with which your child can perform three-digit addition. Download and print the worksheet; then help your child with the first few, and ask them to complete the rest.

217	172	412	383	145	165	173
+124	+228	+399	+681	+455	+226	+237

184	198	438	713	159	558	178
+226	+332	+477	+128	+552	+143	+743

193	137	145	175	347	256	175
+288	+785	+565	+736	+287	+196	+198

FIGURE 2-5: A worksheet for three-digit addition.

Leveraging a Number Line

Teachers introduce kids to number lines in kindergarten to help them learn to count:

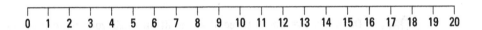

You may have used number lines to help your child solve simple addition problems in earlier grades:

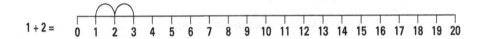

$1 + 2 =$

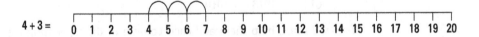

$4 + 3 =$

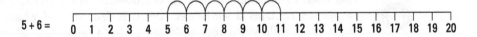

$5 + 6 =$

Using number lines to solve two-digit addition problems

It turns out that new-math techniques leverage number lines to add larger numbers. Ask your child to consider the following expression:

$$25 + 32$$

TIP

You likely don't have a large number line with numbers through 100 readily available. However, you and your child can create one on the fly.

1. Draw a line and mark the first number (25) as shown here:

25

2. Draw three large semicircles to represent the 30 in the second number (32), writing the corresponding number beneath each one, as shown here:

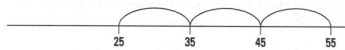

25   35   45   55

3. Draw two small semicircles to represent the 2 in 32, writing the corresponding numbers beneath:

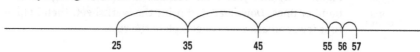

25   35   45   55 56 57

4. Say to your child, "Using a number line, you can quickly solve two-digit addition problems such as 25 plus 32."

Using similar number-line techniques, ask your child to solve the following:

$$55 + 21 \qquad 37 + 21 \qquad 88 + 11$$

Your child should get:

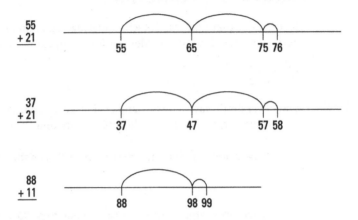

$$
\begin{array}{r} 55 \\ +21 \\ \hline \end{array}
$$

$$
\begin{array}{r} 37 \\ +21 \\ \hline \end{array}
$$

$$
\begin{array}{r} 88 \\ +11 \\ \hline \end{array}
$$

If your child does not, help them draw appropriate large semicircles for the ten's digits and then smaller ones for the one's digits. Make sure that you label the corresponding numbers under each circle.

FIND ONLINE

This book's companion website at www.dummies.com/go/teachingyourkidsnew math6-8fd contains a worksheet (the first few rows of which are shown in Figure 2-6) that your child can use to solve two-digit addition problems using number lines. Download and print the worksheet; then help your child solve the first few, and ask them to complete the rest.

$$
\begin{array}{r} 17 \\ +24 \\ \hline \end{array}
\qquad
\begin{array}{r} 42 \\ +39 \\ \hline \end{array}
\qquad
\begin{array}{r} 45 \\ +45 \\ \hline \end{array}
\qquad
\begin{array}{r} 17 \\ +23 \\ \hline \end{array}
$$

$$
\begin{array}{r} 18 \\ +22 \\ \hline \end{array}
\qquad
\begin{array}{r} 48 \\ +47 \\ \hline \end{array}
\qquad
\begin{array}{r} 19 \\ +52 \\ \hline \end{array}
\qquad
\begin{array}{r} 17 \\ +74 \\ \hline \end{array}
$$

FIGURE 2-6:
Solving two-digit addition using number lines.

$$
\begin{array}{r} 19 \\ +28 \\ \hline \end{array}
\qquad
\begin{array}{r} 14 \\ +56 \\ \hline \end{array}
\qquad
\begin{array}{r} 34 \\ +28 \\ \hline \end{array}
\qquad
\begin{array}{r} 17 \\ +19 \\ \hline \end{array}
$$

Do not move on to three-digit addition using number lines until your child can readily solve two-digit problems.

Using number lines to solve three-digit addition problems

In the previous section, your child used number lines to solve two-digit addition problems. In this section, your child uses a similar approach to solve three-digit problems that consist of hundreds, tens, and ones.

Present the following problem to your child:

$$\begin{array}{r} 317 \\ + 251 \\ \hline \end{array}$$

1. **Have your child draw a line and mark the first number (317) as shown here:**

2. **Ask your child to draw two large semicircles to represent the 200 in the second number (251), writing the corresponding numbers beneath:**

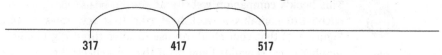

3. **Have your child draw five smaller semicircles to represent the 50 in the second number (251), writing the corresponding numbers beneath:**

4. **Have your child draw one small semicircle to represent the 1 in 251, writing the numbers beneath:**

Present the following problems to your child:

$$\begin{array}{r} 136 \\ + 242 \\ \hline \end{array} \qquad \begin{array}{r} 537 \\ + 341 \\ \hline \end{array} \qquad \begin{array}{r} 414 \\ + 318 \\ \hline \end{array}$$

Using number lines, help your child to solve the problems. They should get:

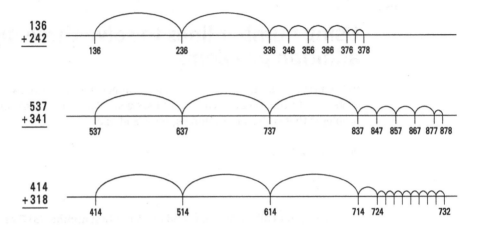

$\begin{array}{r} 136 \\ +242 \\ \hline \end{array}$

$\begin{array}{r} 537 \\ +341 \\ \hline \end{array}$

$\begin{array}{r} 414 \\ +318 \\ \hline \end{array}$

If they do not, take time to represent the hundreds, tens, and ones in the second number of each addition using sized semicircles, writing the corresponding numbers beneath.

FIND ONLINE

This book's companion website at www.dummies.com/go/teachingyourkidsnewmath6-8fd contains a worksheet (the first few rows of which are shown in Figure 2-7) that your child can use to solve three-digit addition problems using number lines. Download and print the worksheet; then help your child solve the first few and ask them to complete the rest.

$\begin{array}{r} 171 \\ +242 \\ \hline \end{array}$	$\begin{array}{r} 422 \\ +392 \\ \hline \end{array}$	$\begin{array}{r} 452 \\ +451 \\ \hline \end{array}$	$\begin{array}{r} 171 \\ +232 \\ \hline \end{array}$
$\begin{array}{r} 183 \\ +223 \\ \hline \end{array}$	$\begin{array}{r} 481 \\ +472 \\ \hline \end{array}$	$\begin{array}{r} 192 \\ +522 \\ \hline \end{array}$	$\begin{array}{r} 127 \\ +724 \\ \hline \end{array}$
$\begin{array}{r} 192 \\ +282 \\ \hline \end{array}$	$\begin{array}{r} 141 \\ +516 \\ \hline \end{array}$	$\begin{array}{r} 314 \\ +228 \\ \hline \end{array}$	$\begin{array}{r} 217 \\ +139 \\ \hline \end{array}$

FIGURE 2-7: A worksheet for solving three-digit addition problems using number lines.

Working Out Addition Word Problems

Word problems get a bad rap. Many of us remember difficult word problems from our high-school math days. These problems exist to bring real-world scenarios into math problems. In other words, word problems provide examples that address the question, "Why do I need to learn this?" In the sections that follow, your child moves from easy to more challenging word problems that require addition.

Getting Started with Simple Word Problems

The first few word problems require your child to solve single-digit math operations. The math operations are easy; the challenge is in determining what the problem is asking.

Regardless of the word problem, your child should start the process by determining the question being asked.

Example: Three ducks are swimming in a lake. Four more ducks land and begin to swim. How many ducks are in the lake?

In this case, the problem specifically asks, "How many ducks are in the lake?"

The second step in solving word problems is determining the type of math operation to perform, such as addition or subtraction. In this case, you are solving the following addition problem:

$$\begin{array}{r} 3 \\ +\ 4 \\ \hline \end{array}$$

As the word problems become more difficult, your child may find it helpful to draw a picture. In this case, they might create the following:

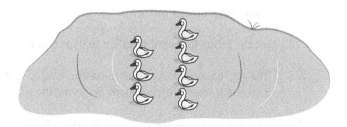

Present the following word problem to your child, asking the two questions and optionally drawing a picture:

Bill, Fred, and Javier went fishing. Bill caught 2 fish, Fred 3, and Javier 1. How many fish did the boys catch?

Your child should get the following:

$$
\begin{array}{r}
2 \\
+\ 3 \\
+\ 1 \\
\hline
6
\end{array}
$$

Present the following word problem to your child:

Martha, Shaunae, and Holly each read books over the weekend. Martha read 5, Shaunae read 6, and Holly 3. How many books did Martha and Holly read?

Okay, this problem illustrates why word problems get a bad rap. The problem states the number of books three girls read, but then asks your child to solve only for the number of books read by two girls: Martha and Holly. Your child should get:

$$
\begin{array}{r}
5 \\
+\ 3 \\
\hline
8
\end{array}
$$

Solving two-digit addition word problems

In this section, the word problems require two-digit addition. The math is slightly harder, but the steps to solve the problems remain the same:

1. **Determine the question being asked.**

2. **Determine the type of math operation.**

3. **Optionally, draw a picture.**

Present the following word problem to your child:

The girls' basketball team won the championship. Donna scored 10 points and her sister Beth scored 16. How many points did the sisters score?

Your child should identify and solve for the following:

$$
\begin{array}{r}
10 \\
+\ 16 \\
\hline
26
\end{array}
$$

Present the following word problem to your child:

> Daniella's mom went grocery shopping. She bought pizza for $10, soda for $6, a cake for $12, and ice cream for $4. How much did her groceries cost?

Your child should identify and solve for the following:

```
   10
 + 6
 +12
 + 4
 ----
   32
```

Solving three-digit addition word problems

In this section, your child solves word problems that require three-digit addition. As the numbers get bigger, the problems can seem harder. Remember, however, that the three steps to solving a word problem do not change.

Present the following word problem to your child:

> The school's two-member bike club rode every day for a month. Bill rode 150 miles and Sasha rode 175. How many miles did the two ride?

Your child should identify and solve the following:

```
  150
 +175
 ----
  325
```

Present the following word problem to your child:

> Nancy's family bought a puppy and a kitten. The puppy cost $200 and the kitten $175. How much did Nancy's family spend on the pets?

Your child should identify and solve the following:

```
  200
 +175
 ----
  375
```

IN THIS CHAPTER

» **Helping your child to do subtraction problems in their head**

» **Using boxes to assist with borrowing**

» **Borrowing without boxes**

» **Subtracting using number lines**

» **Solving subtraction word problems**

Chapter **3**

Subtracting Any Knowledge Gaps

S ubtraction, in the simplest sense, is the opposite of addition. However, psychologically, it always seems better to be adding something than taking it away!

That said, this chapter reviews subtraction. As you did for addition, you start with flashcards that help your child perform key subtraction operations in their head. Not only does this let them show off their math skills, but it also lays a solid foundation for multi-digit subtraction operations.

You then look at old-school subtraction — the subtraction you learned with borrowing (which teachers today often call *regrouping*) — followed by new-math subtraction using number lines, which eliminates the need to regroup. Finally, you examine some word problems that require your child to solve using subtraction.

Solving Subtraction Problems in Their Heads

In Chapter 2, your child worked on performing key addition operations in their head. In this section, they perform similar operations, but this time using subtraction. Again, being able to perform such operations mentally will significantly improve their ability to perform multi-digit subtraction.

Using your 3x5 index cards, create the following flashcards. (Or, alternatively, you can purchase a pack of flashcards.)

0	1	2	3	4	5	6	7	8	9	10
0 −0	1 −0	2 −0	3 −0	4 −0	5 −0	6 −0	7 −0	8 −0	9 −0	10 −0
	1 −1	2 −1	3 −1	4 −1	5 −1	6 −1	7 −1	8 −1	9 −1	10 −1
		2 −2	3 −2	4 −2	5 −2	6 −2	7 −2	8 −2	9 −2	10 −2
			3 −3	4 −3	5 −3	6 −3	7 −3	8 −3	9 −3	10 −3
				4 −4	5 −4	6 −4	7 −4	8 −4	9 −4	10 −4
					5 −5	6 −5	7 −5	8 −5	9 −5	10 −5
						6 −6	7 −6	8 −6	9 −6	10 −6
							7 −7	8 −7	9 −7	10 −7
								8 −8	9 −8	10 −8
									9 −9	10 −9
										10 −10

If you want your child to practice the flashcards on their own, write the correct answer on the back of each card so they can quickly check their result.

These flashcards all have results that are in the range of 0 to 10. By mastering these flashcards, you lay the foundation that your child will use when they perform multi-digit addition.

Mastering timed subtraction tests

Your goal in practicing subtraction flashcards is for your child to quickly and accurately solve key subtraction problems in their head. After your child has mastered the flashcards, have them practice timed worksheets similar to that shown in Figure 3-1, which I've provided on this book's companion website at www.dummies.com/go/teachingyourkidsnewmath6-8fd.

Ideally, your child should accurately complete the worksheet in 5 minutes or less. When they can do so, you can move on.

Timed tests can be stressful, so try to keep the process fun. Be positive about the problems your child gets right. If they miss several problems, that's your sign that they should spend more time with the flashcards.

Your child's ability to solve common subtraction problems in their head will build their confidence and lay the foundation for future math success.

Counting cards — so to speak

After your child can subtract numbers through 10 in their head, take out a deck of playing cards, removing the joker. Explain to your child that in some card games, the ace can count as 1 or 11 — for this game, it counts as 1. Explain to your child that each face card counts as 10.

Spread the cards face down on a table (or on the floor) and ask your child to turn over 2 cards. Ask your child to subtract the smaller card from the larger card in their head, saying the correct answer out loud. If your child is correct, they get to keep the cards; otherwise, you get to keep the cards. When all the cards have been turned over, the player with the most cards wins.

If your child easily subtracts the cards in the deck, you can advance the game by having your child turn over two cards and seeing which of you can say the answer faster. Whoever correctly says the answer first gets to keep the card.

$$\begin{array}{ccccccc}
4 & 2 & 4 & 8 & 4 & 8 & 7 \\
-2 & -2 & -3 & -6 & -4 & -6 & -3 \\
\hline
\end{array}$$

$$\begin{array}{ccccccc}
8 & 9 & 5 & 7 & 10 & 5 & 7 \\
-2 & -3 & -4 & -6 & -9 & -0 & -7 \\
\hline
\end{array}$$

$$\begin{array}{ccccccc}
9 & 9 & 7 & 8 & 10 & 9 & 10 \\
-8 & -7 & -5 & -7 & -8 & -9 & -9 \\
\hline
\end{array}$$

$$\begin{array}{ccccccc}
7 & 9 & 10 & 6 & 7 & 8 & 10 \\
-5 & -8 & -10 & -6 & -5 & -8 & -8 \\
\hline
\end{array}$$

$$\begin{array}{ccccccc}
1 & 2 & 5 & 8 & 8 & 10 & 8 \\
-1 & -0 & -3 & -7 & -6 & -9 & -3 \\
\hline
\end{array}$$

$$\begin{array}{ccccccc}
7 & 8 & 6 & 10 & 10 & 7 & 8 \\
-5 & -7 & -1 & -6 & -4 & -3 & -7 \\
\hline
\end{array}$$

FIGURE 3-1:
A timed
subtraction
worksheet.

$$\begin{array}{ccccccc}
10 & 10 & 4 & 6 & 3 & 10 & 7 \\
-10 & -2 & -3 & -5 & -3 & -4 & -3 \\
\hline
\end{array}$$

Taking Flashcards to the Next Level

It's important for their future math success that your child can quickly subtract numbers in their head. That said, you don't have to stop your flashcards at 10. You might, for example, add the following flashcards to your practice deck:

11	11	11	11	11	11	11	11	11	11
−1	−2	−3	−4	−5	−6	−7	−8	−9	−10

After your child has mastered the cards, you can add the following:

12	12	12	12	12	12	12	12	12	12
−1	−2	−3	−4	−5	−6	−7	−8	−9	−10

You can repeat this process of adding cards through 100.

REMEMBER

The ability to subtract numbers through 100 in their head will take time and practice. Do not add additional cards to your child's flashcards until they have mastered the current deck. It may take your child weeks or even months of flashcard practice to master numbers through 100. The result, however, will be worth the effort.

Subtraction with Regrouping

As I mention in Chapter 2, *regrouping* is a ten-cent word for carrying and borrowing — the old-school way most of us learned to add and subtract. In this section, your child will practice subtraction with regrouping (aka borrowing). Present the following problems to your child:

$$\begin{array}{r} 13 \\ -5 \\ \hline \end{array} \qquad \begin{array}{r} 17 \\ -8 \\ \hline \end{array} \qquad \begin{array}{r} 12 \\ -3 \\ \hline \end{array} \qquad \begin{array}{r} 14 \\ -7 \\ \hline \end{array}$$

In the first problem, your child can't subtract 5 from 3, but they can subtract 5 from 13. Likewise, in the second problem, your child can't subtract 8 from 7, but they can subtract 8 from 17.

Have your child solve the problems. They should get:

$$\begin{array}{r} 13 \\ -5 \\ \hline 8 \end{array} \qquad \begin{array}{r} 17 \\ -8 \\ \hline 9 \end{array} \qquad \begin{array}{r} 12 \\ -3 \\ \hline 9 \end{array} \qquad \begin{array}{r} 14 \\ -7 \\ \hline 7 \end{array}$$

If your child does not get the right answers, it's a signal that they should spend more time with the subtraction flashcards.

Borrowing with boxes

When you use borrowing to solve a subtraction problem, you rewrite the number from which you borrowed, as shown here:

$$\begin{array}{r} {}^{2}\cancel{3}^{1}1 \\ -14 \\ \hline 17 \end{array}$$

To help your child master the borrowing process, this section presents subtraction problems with boxes into which your child can write the new numbers. As your child borrows from the tens column, remind them they are bring over the value 10.

$$\begin{array}{r} \square \\ 3^{\square}1 \\ -14 \\ \hline \end{array} \qquad \begin{array}{r} \boxed{2} \\ \cancel{3}^{\square}1 \\ -14 \\ \hline 17 \end{array}$$

Present the following problem to your child:

$$\begin{array}{r} \square \\ 3^{\square}5 \\ -18 \\ \hline \end{array}$$

1. **Explain to your child that they cannot subtract 8 from 5, and so they must borrow 10 from the 30.**

2. **Have your child perform the borrow operation, crossing out the 3 and updating the boxes:**

$$\begin{array}{r} \boxed{2} \\ \cancel{3}^{\square}5 \\ -18 \\ \hline \end{array}$$

3. **Perform the subtraction:**

$$\begin{array}{r} \boxed{2} \\ \cancel{3}^{\square}5 \\ -18 \\ \hline 17 \end{array}$$

Ask your child to solve the following problems, assisting as necessary:

$$\begin{array}{r} \square \\ 4\,\square\,4 \\ -2\ 7 \end{array} \qquad \begin{array}{r} \square \\ 3\,\square\,5 \\ -1\ 9 \end{array} \qquad \begin{array}{r} \square \\ 2\,\square\,3 \\ -1\ 4 \end{array}$$

Your child should get the following:

$$\begin{array}{r} \boxed{3} \\ \cancel{4}\,\boxed{\square}\,4 \\ -2\ 7 \\ \hline 1\ 7 \end{array} \qquad \begin{array}{r} \boxed{2} \\ \cancel{3}\,\boxed{\square}\,5 \\ -1\ 9 \\ \hline 1\ 6 \end{array} \qquad \begin{array}{r} \boxed{1} \\ \cancel{2}\,\boxed{\square}\,3 \\ -1\ 4 \\ \hline 9 \end{array}$$

If they do not, walk them through the borrowing process, using the boxes to write the borrowed number and updated tens digit.

FIND
ONLINE

This book's companion website at www.dummies.com/go/teachingyourkidsnew math6-8fd contains a worksheet (the first few rows of which are shown in Figure 3-2) that your child can use to perform borrowing actions using boxes. Download and print the worksheet. Help your child solve the first few and then ask them to complete the rest.

$$\begin{array}{r} \square \\ 7\,\square\,3 \\ -2\ 7 \\ \hline \end{array} \qquad \begin{array}{r} \square \\ 8\,\square\,1 \\ -1\ 5 \\ \hline \end{array} \qquad \begin{array}{r} \square \\ 2\,\square\,1 \\ -1\ 6 \\ \hline \end{array} \qquad \begin{array}{r} \square \\ 3\,\square\,3 \\ -1\ 7 \\ \hline \end{array}$$

FIGURE 3-2:
A subtraction
worksheet with
boxes to assist in
the borrowing
process.

$$\begin{array}{r} \square \\ 5\,\square\,1 \\ -3\ 7 \\ \hline \end{array} \qquad \begin{array}{r} \square \\ 4\,\square\,2 \\ -2\ 4 \\ \hline \end{array} \qquad \begin{array}{r} \square \\ 1\,\square\,3 \\ -\ \ 8 \\ \hline \end{array} \qquad \begin{array}{r} \square \\ 2\,\square\,5 \\ -1\ 6 \\ \hline \end{array}$$

Using boxes to subtract three-digit numbers

The previous section covers two-digit subtraction using boxes to assist with borrowing. In this section, your child can use a similar process for larger three-digit numbers:

$$\begin{array}{r} \square\ \square \\ 3\ \square\,1\ \square\,4 \\ -1\ \ 7\ \ 6 \end{array}$$

Present the following problem to your child:

$$
\begin{array}{r}
\boxed{2}\ \boxed{1}\\
\cancel{3}\ ^{\boxed{1}}\cancel{2}\ ^{\boxed{1}}2\\
-1\ \ \ 7\ \ \ 6\\
\hline
1\ \ \ 4\ \ \ 6
\end{array}
$$

1. **Explain to your child that they cannot subtract 6 from 4, so they must borrow 10. Update the boxes and perform the subtraction.**

$$
\begin{array}{r}
\ \boxed{\ }\ \boxed{0}\\
3\ ^{\boxed{\ }}\cancel{1}\ ^{\boxed{\ }}4\\
-1\ \ \ 7\ \ \ 6\\
\hline
8
\end{array}
$$

2. **Explain to your child that they cannot subtract 7 from 0, so they must again borrow. Update the boxes and complete the subtraction:**

$$
\begin{array}{r}
\boxed{2}\ \boxed{0}\\
\cancel{3}\ ^{\boxed{10}}\cancel{1}\ ^{\boxed{\ }}4\\
-1\ \ \ 7\ \ \ 6\\
\hline
3\ \ \ 8
\end{array}
$$

3. **Have your child perform the final subtraction:**

$$
\begin{array}{r}
\boxed{2}\ \boxed{0}\\
\cancel{3}\ ^{\boxed{10}}\cancel{1}\ ^{\boxed{\ }}4\\
-1\ \ \ 7\ \ \ 6\\
\hline
1\ \ \ 3\ \ \ 8
\end{array}
$$

4. **Ask your child to solve the following problem, assisting as necessary:**

$$
\begin{array}{r}
\boxed{\ }\ \boxed{\ }\\
3\ ^{\boxed{\ }}1\ ^{\boxed{\ }}3\\
-1\ \ \ 6\ \ \ 7\\
\end{array}
$$

Your child should get

$$
\begin{array}{r}
\boxed{2}\ \boxed{0}\\
\cancel{3}\ ^{\boxed{10}}\cancel{1}\ ^{\boxed{\ }}3\\
-1\ \ \ 6\ \ \ 7\\
\hline
1\ \ \ 4\ \ \ 6
\end{array}
$$

FIND ONLINE

This book's companion website at www.dummies.com/go/teachingyourkidsnew math6-8fd contains a worksheet (the first few rows of which are shown in Figure 3-3) that your child can use to complete three-digit subtractions using boxes to assist with borrowing. Download and print the worksheet. Help your child solve the first few and then ask them to complete the rest.

$$\square\ \square$$
$$3\ ^{\square}1\ ^{\square}7$$
$$-\ 2\ 5\ 8$$

$$\square\ \square$$
$$1\ ^{\square}5\ ^{\square}7$$
$$-\ 1\ 3\ 8$$

$$\square\ \square$$
$$2\ ^{\square}2\ ^{\square}2$$
$$-\ 1\ 5\ 9$$

FIGURE 3-3:
A worksheet with boxes to assist with borrowing.

$$\square\ \square$$
$$5\ ^{\square}0\ ^{\square}1$$
$$-\ 3\ 5\ 7$$

$$\square\ \square$$
$$4\ ^{\square}0\ ^{\square}0$$
$$-\ 2\ 9\ 7$$

$$\square\ \square$$
$$3\ ^{\square}1\ ^{\square}3$$
$$-\ 1\ 5\ 7$$

Removing the boxes

The previous subtraction problems use boxes to help your child subtract multi-digit numbers that require regrouping (I still like the term *borrowing*). In this section, your child performs the same operations, but this time, without the boxes.

Present the following problems to your child:

$$\begin{array}{ccc} 31 & 52 & 77 \\ -17 & -25 & -28 \end{array}$$

TIP

Remind your child: "When you borrow, you must cross out the current tens digit and rewrite the updated number, as you did using the boxes."

Your child should get

$$\begin{array}{ccc} ^{2} & ^{4} & ^{6} \\ \cancel{3}\,^{1}1 & \cancel{5}\,^{1}2 & \cancel{7}\,^{1}7 \\ -1\,7 & -2\,5 & -2\,8 \\ \hline 1\,4 & 2\,7 & 4\,9 \end{array}$$

If they do not, you may want them to repeat the two-digit subtraction worksheet that leveraged the boxes.

FIND ONLINE

This book's companion website at www.dummies.com/go/teachingyourkidsnew math6-8fd contains a worksheet (the first few rows of which are shown in Figure 3-4) that your child can use to perform two-digit subtraction with borrowing. Download and print the worksheet. Help your child solve the first few and then ask them to complete the rest.

REMEMBER

Do not move onto three-digit subtraction until your child masters the two-digit problems.

| 41 | 32 | 42 | 83 | 45 | 65 | 32 |
| −24 | −28 | −39 | −68 | −45 | −26 | −23 |

| 28 | 31 | 48 | 73 | 43 | 58 | 37 |
| −19 | −23 | −19 | −18 | −34 | −19 | −18 |

| 12 | 43 | 74 | 17 | 34 | 25 | 37 |
| − 8 | −28 | −56 | − 9 | −18 | −19 | −19 |

FIGURE 3-4:
A worksheet for
performing
two-digit
subtraction with
borrowing.

Present the following problems to your child, assisting as necessary:

$$\begin{array}{ccc} 322 & 555 & 417 \\ -133 & -366 & -278 \end{array}$$

Your child should get the following:

$$\begin{array}{ccc} {}^{2}\;\;\;\;\; & {}^{4}\;\;\;\;\; & {}^{3}\;\;\;\;\; \\ \cancel{3}{}^{11}\cancel{2}{}^{1}2 & \cancel{5}{}^{14}\cancel{5}{}^{1}5 & \cancel{4}{}^{10}\cancel{1}{}^{1}7 \\ -1\;\;3\;\;3 & -3\;\;6\;\;6 & -2\;\;7\;\;8 \\ \hline 1\;\;8\;\;9 & 1\;\;8\;\;9 & 1\;\;3\;\;9 \end{array}$$

If they do not, you may want to review the three-digit subtraction worksheet that uses boxes to assist with borrowing.

FIND ONLINE

This book's companion website at www.dummies.com/go/teachingyourkidsnew math6–8fd contains a worksheet (the first few rows of which are shown in Figure 3-5) that your child can use to perform three-digit subtraction. Download and print the worksheet. Help your child with the first few and then ask them to complete the rest.

| 412 | 323 | 422 | 833 | 455 | 651 | 332 |
| −243 | −282 | −395 | −684 | −428 | −262 | −233 |

| 283 | 314 | 485 | 733 | 434 | 586 | 373 |
| −194 | −235 | −179 | −184 | −346 | −198 | −184 |

| 124 | 435 | 745 | 173 | 345 | 255 | 375 |
| − 85 | −287 | −569 | − 95 | −188 | −199 | −199 |

FIGURE 3-5:
A worksheet for
three-digit
subtraction.

Subtracting Using a Number Line

Chapter 2 covers the use of number lines to perform new-math addition techniques. In this section, your child gets to use a number line to practice subtraction.

You may have used number lines to help your child solve simple subtraction problems, such as

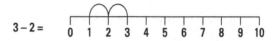

$3 - 2 =$

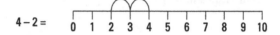

$4 - 2 =$

$7 - 5 =$

Using number lines to practice two-digit subtraction

Just as you found with addition, new-math techniques use number lines for two-digit subtraction. Ask your child to consider the following expression:

$$\begin{array}{r} 33 \\ -24 \\ \hline \end{array}$$

1. Draw a line and mark the first number (33) as shown here:

33

2. Draw two large semicircles to represent the 20 in the second number (24), writing the corresponding number beneath each one, as shown here:

13   23   33

3. Draw four small semicircles to represent the 4 in 24, writing the corresponding numbers beneath:

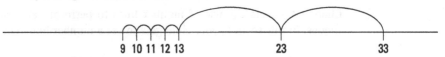

9 10 11 12 13   23   33

4. Say to your child, "Using a number line, you can quickly solve two-digit subtraction problems such as 33 minus 24."

Using similar number-line techniques, ask your child to solve the following:

$$\begin{array}{ccc} 51 & 67 & 83 \\ -24 & -48 & -24 \end{array}$$

Your child should get

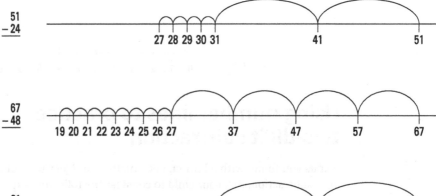

$\begin{array}{c} 51 \\ -24 \end{array}$ 27 28 29 30 31   41   51

$\begin{array}{c} 67 \\ -48 \end{array}$ 19 20 21 22 23 24 25 26 27   37   47   57   67

$\begin{array}{c} 81 \\ -24 \end{array}$ 57 58 59 60 61   71   81

TIP

If your child does not, help them draw appropriate large semicircles for the tens digits and then smaller ones for the ones digits. Make sure that you label the corresponding numbers under each circle.

FIND ONLINE

This book's companion website at www.dummies.com/go/teachingyourkidsnew math6-8fd contains a worksheet (the first few rows of which are shown in Figure 3-6) that your child can use to solve two-digit subtraction problems using number lines. Download and print the worksheet. Help your child solve the first few and then ask them to complete the rest.

REMEMBER

Do not move on to three-digit subtraction using number lines until your child can readily solve two-digit problems.

37	42	45	37
− 24	− 39	− 45	− 23

48	48	79	97
− 22	− 47	− 52	− 74

FIGURE 3-6:
Solving two-digit subtraction using number lines.

39	74	34	27
− 28	− 56	− 28	− 19

Using number lines to solve three-digit subtraction problems

In the previous section, your child used number lines to solve two-digit subtraction problems. In this section, your child uses a similar approach to solve three-digit problems that consists of hundreds, tens, and ones.

Present the following problem to your child:

313
−134

1. Have your child draw a line and mark the first number (313), as shown here:

317

2. Ask your child to draw one large semicircle to represent the 100 in the second number (134), writing the corresponding numbers beneath:

217   317

3. Have your child draw three smaller semicircles to represent the 30 in the second number (134), writing the corresponding numbers beneath:

187   197   207   217   317

4. Have your child draw four small semicircles to represent the 4 in 134, writing the numbers beneath:

183 184 185 186 187   197   207   217   317

Present the following problems to your child:

$$\begin{array}{ccc} 322 & 617 & 801 \\ -143 & -158 & -222 \end{array}$$

Using number lines, help your child solve the problems. They should get

$\begin{array}{c} 322 \\ -143 \end{array}$

179 180 181 182   192   202   212   222   322

$\begin{array}{c} 617 \\ -158 \end{array}$

449   457   467   477   487   497   507   517   617

$\begin{array}{c} 801 \\ -222 \end{array}$

579   581   591   601   701   801

TIP

If they do not, take time to represent the hundreds, tens, and ones in the second number of each addition problem using sized semicircles, and writing the corresponding numbers beneath.

FIND ONLINE

This book's companion website at www.dummies.com/go/teachingyourkidsnew math6–8fd contains a worksheet (the first few rows of which are shown in Figure 3-7) that your child can use to solve three-digit subtraction problems using number lines. Download and print the worksheet. Help your child solve the first few and then ask them to complete the rest.

371	422	452	471
−242	−392	−451	−232

383	481	692	827
−223	−472	−522	−724

392	541	314	217
−282	−516	−228	−139

Working Out Subtraction Word Problems

You may be thinking, "Great! More word problems!" Remember that word problems exist to present real-world math scenarios. In this section, your child solves word problems that require subtraction.

REMEMBER

Keep in mind the three steps to solving word problems:

1. **Determine the question being asked.**

2. **Determine the math operation, such as addition or subtraction.**

3. **Optionally, draw a picture to illustrate the problem.**

Getting started with subtraction word problems

The first word problems in this section require your child to use single-digit subtraction. Present the following word problem to your child, asking the two questions and optionally drawing a picture:

> Seven ducks are swimming in a lake. Four ducks fly away. How many ducks remain in the lake?

Your child should get the following:

$$\begin{array}{r} 7 \\ -4 \\ \hline 3 \end{array}$$

If your child does not, you can draw a picture to help them, as shown here:

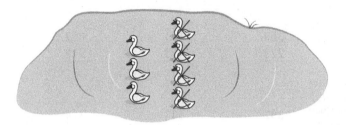

Present the following word problem to your child:

Mary and Shawn are selling boxes of cookies at a stand. They started with 10 boxes. Mary sold 3 boxes and Shawn sold 4. How many boxes do they have left?

Your child should get:

$$
\begin{array}{r}
10 \\
-3 \\
\underline{-4} \\
3
\end{array}
$$

If they do not, you can again draw a picture:

Solving two-digit subtraction word problems

In this section, the word problems require two-digit subtraction. The math is slightly harder, but the steps to solve the problems remain the same:

1. **Determine the question being asked.**

2. **Determine the type of math operation.**

3. **Optionally, draw a picture.**

Present the following word problem to your child:

> Javier went shopping for jeans. Javier had $40 and bought jeans costing $24. How much money does Javier have left?

Your child should identify the starting amount and the amount being subtracted, and then solve the problem:

$$
\begin{array}{r}
40 \\
-24 \\
\hline
16
\end{array}
$$

If they do not, help them identify the starting amount ($40) and the amount being subtracted ($24).

Present the following word problem to your child:

> AJ had one hour (60 minutes) before bedtime. He spent 10 minutes on flashcards and 20 minutes reading. How much spare time did AJ have left?

Your child should identify the starting amount and the amounts being subtracted, and then solve the problem:

$$
\begin{array}{r}
60 \\
-10 \\
-20 \\
\hline
30
\end{array}
$$

If they do not, try drawing the following picture:

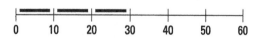

Solving three-digit subtraction word problems

In this section, your child solves word problems that require three-digit subtraction. As the numbers get bigger, the problems can seem harder. Remember, however, that the three steps to solving a word problem do not change.

Present the following word problem to your child:

> Angela wants to buy a new bike that costs $250. She has saved $120. How much more does Angela need to save before she can buy the new bike?

Your child should identify the starting amount and the amount being subtracted, and then solve the problem:

$$\begin{array}{r} 250 \\ -120 \\ \hline 130 \end{array}$$

If they do not, help them structure the problem by writing the starting value (250) and the amount they should subtract from it (120).

Present the following word problem to your child:

> Alonzo and Terry want to ride their bikes 500 miles over spring break. On the first day, they ride 125 miles. How much more must they ride to reach their goal?

Your child should identify the starting amount and the amount being subtracted, and then solve the problem:

$$\begin{array}{r} 500 \\ -125 \\ \hline 375 \end{array}$$

Chapter **4**

Multiplying Their Multiplication Skills

Multiplication is the process of multiplying! There, that was easy! If that wasn't helpful, you can think of multiplication as a math technique that helps you quickly perform repeated addition. In other words, 5×5 represents $5 + 5 + 5 + 5 + 5$ and 10×5 represents $10 + 10 + 10 + 10 + 10$.

This chapter is about multiplication. As you might have guessed, I start the process with flashcards that cover key multiplication operations from 1×1 through 10×10. The ability of your child to quickly (and accurately) solve common multiplication operations in their head will be a great life skill.

After your child masters a few timed multiplication tests, you look at old-school multiplication techniques — the way you likely learned to multiply. Then, you get introduced to some new math with box multiplication.

And, as you might have also guessed, you get to knock out some word problems that require your child to multiply.

So, get ready to go forth and multiply.

Break Out the Flashcards: Brushing Up on Multiplication Factors

The key to your child's success with multi-digit multiplication is being able to quickly and accurately solve the multiplication factors through 100. The best way to accomplish that skill is by using flashcards.

Using your 3x5 index cards, create the following flashcards:

0 ×0	0 ×1	0 ×2	0 ×3	0 ×4	0 ×5	0 ×6	0 ×7	0 ×8	0 ×9	0 ×10
1 ×0	1 ×1	1 ×2	1 ×3	1 ×4	1 ×5	1 ×6	1 ×7	1 ×8	1 ×9	1 ×10
2 ×0	2 ×1	2 ×2	2 ×3	2 ×4	2 ×5	2 ×6	2 ×7	2 ×8	2 ×9	2 ×10
3 ×0	3 ×1	3 ×2	3 ×3	3 ×4	3 ×5	3 ×6	3 ×7	3 ×8	3 ×9	3 ×10
4 ×0	4 ×1	4 ×2	4 ×3	4 ×4	4 ×5	4 ×6	4 ×7	4 ×8	4 ×9	4 ×10
5 ×0	5 ×1	5 ×2	5 ×3	5 ×4	5 ×5	5 ×6	5 ×7	5 ×8	5 ×9	5 ×10
6 ×0	6 ×1	6 ×2	6 ×3	6 ×4	6 ×5	6 ×6	6 ×7	6 ×8	6 ×9	6 ×10
7 ×0	7 ×1	7 ×2	7 ×3	7 ×4	7 ×5	7 ×6	7 ×7	7 ×8	7 ×9	7 ×10
8 ×0	8 ×1	8 ×2	8 ×3	8 ×4	8 ×5	8 ×6	8 ×7	8 ×8	8 ×9	8 ×10
9 ×0	9 ×1	9 ×2	9 ×3	9 ×4	9 ×5	9 ×6	9 ×7	9 ×8	9 ×9	9 ×10
10 ×0	10 ×1	10 ×2	10 ×3	10 ×4	10 ×5	10 ×6	10 ×7	10 ×8	10 ×9	10 ×10

TIP

If you want your child to practice the flashcards on their own, write the correct answer on the back of each card so they can quickly check their result.

These flashcards all have results that are in the range of 0 to 100. By mastering these flashcards, you lay the foundation that your child will use when they perform multi-digit multiplication.

TIP

You should find that, just as practicing the addition and subtraction flashcards made your child faster and more accurate, the same is true for multiplication. Plan on spending 10 to 15 minutes a day practicing the multiplication flashcards until your child masters them.

Mastering timed multiplication tests

FIND
ONLINE

Your goal in practicing multiplication flashcards is for your child to quickly and accurately solve key multiplication problems in their head. After your child has mastered the flashcards, have them practice timed worksheets similar to that shown in Figure 4-1, which I have provided on this book's companion website at www.dummies.com/go/teachingyourkidsnewmath6-8fd.

Ideally, your child should accurately complete the worksheet in 5 minutes or less. When they can do so, you can move on.

TIP

Timed tests can be stressful, so try to keep the process fun. Be positive about the problems your child gets right. If they miss several problems, that's your sign that they should spend more time with the flashcards.

REMEMBER

Your child's ability to solve common multiplication problems in their head will build their confidence and lay the foundation for future math success.

Back to counting cards for more multiplication fun

Once your child can multiply numbers through 100 in their head, take out a deck of playing cards, removing the joker. Explain to your child that in some card games, the ace can count as 1 or 11 — for this game, it counts as 1. Explain to your child that each face card counts as 10.

Spread the cards face down on a table (or on the floor) and ask your child to turn over 2 cards. Ask your child to multiply the cards in their head, saying the correct answer out loud. If your child is correct, they get to keep the cards; otherwise, you get to keep the cards. When all the cards have been turned over, the player with the most cards wins.

1 ×2	2 ×2	4 ×3	3 ×6	4 ×4	5 ×6	7 ×3
8 ×2	9 ×3	4 ×4	7 ×6	9 ×9	5 ×0	7 ×7
1 ×8	9 ×7	4 ×5	7 ×7	4 ×8	5 ×9	7 ×9
1 ×5	4 ×9	4 ×10	6 ×6	5 ×5	5 ×8	7 ×8
1 ×1	2 ×0	5 ×3	4 ×7	6 ×8	1 ×9	8 ×3
6 ×9	9 ×7	6 ×1	10 ×6	10 ×4	9 ×3	10 ×7
10 ×10	10 ×2	4 ×10	1 ×8	3 ×3	15 ×4	7 ×3

FIGURE 4-1: A timed multiplication worksheet.

If your child easily adds the cards in the deck, you can advance the game by having your child turn over 2 cards and seeing which of you can say the answer faster. Whoever correctly says the answer first gets to keep the cards.

Becoming One of "Those" People Who Multiplies in their Head

It's important for their future math success that your child can quickly multiply numbers through 100 in their head. That said, you don't have to stop your flash-cards at 100. You might, for example, add the following flashcards to your practice deck:

11 ×1	11 ×2	11 ×3	11 ×4	11 ×5	11 ×6	11 ×7	11 ×8	11 ×9	11 ×10

After your child has mastered these cards, you can add the following:

12 ×1	12 ×2	12 ×3	12 ×4	12 ×5	12 ×6	12 ×7	12 ×8	12 ×9	12 ×10

You can repeat this process of adding cards through 20.

REMEMBER

The ability to multiply numbers through 200 (10x20) in their head will take time and practice. Do not add additional cards to your child's flashcards until they have mastered the current deck. It may take your child weeks or even months of flash-card practice to master numbers through 200. The result, however, will be worth the effort.

Old-School Multi-digit Multiplication

Before looking at new-math multiplication, this section spends some time on doing multi-digit multiplication the old-school way using both two-digit and three-digit numbers.

Solving two-digit multiplication problems

Present the following problem to your child:

```
  17
×13
```

1. Have your child start by multiplying 3x7. They should get 21. Have them write down the 1 and carry the 2 to the tens column:

```
  2
  17
×13
   1
```

2. Have your child multiply 3x1, which is 3. They can then add the 2, getting 5, which they write down:

```
  2
  17
×13
  51
```

3. Explain to your child, "When you move to the tens column, you must first write down a 0 in the ones column:"

```
  2
  17
×13
  51
   0
```

4. Have your child multiply 1x7, writing down the result:

```
  2
  17
×13
  51
  70
```

5. Have your child multiply 1x1, writing down the result:

```
  2
  17
×13
  51
 170
```

6. Have your child add up the two rows to get the final result:

```
  2
  17
×13
  51
 170
 221
```

Repeat these steps with your child to solve the following problems:

$$\begin{array}{r} 25 \\ \times 13 \\ \hline \end{array} \qquad \begin{array}{r} 19 \\ \times 18 \\ \hline \end{array} \qquad \begin{array}{r} 22 \\ \times 33 \\ \hline \end{array}$$

Your child should get the following:

$$\begin{array}{r} 25 \\ \times 13 \\ \hline 75 \\ 250 \\ \hline 325 \end{array} \qquad \begin{array}{r} 19 \\ \times 18 \\ \hline 152 \\ 190 \\ \hline 342 \end{array} \qquad \begin{array}{r} 22 \\ \times 33 \\ \hline 66 \\ 660 \\ \hline 726 \end{array}$$

If your child does not, or is not familiar with the old-school multiplication process, consider strengthening their math foundation with the book *Teaching Your Kids New Math, K-5 For Dummies*, also written by me and published by Wiley.

TIP

Your child should correctly and quickly multiply the two digits in each column. If they have problems, that's a signal that you should spend more time practicing with the multiplication flashcards.

FIND ONLINE

This book's companion website at www.dummies.com/go/teachingyourkidsnew math6-8fd contains a worksheet (the first few rows of which are shown in Figure 4-2) that your child can use to practice two-digit multiplication. Download and print the worksheet. Help your child solve the first few and then ask them to complete the rest.

$$\begin{array}{r} 17 \\ \times 24 \\ \hline \end{array} \quad \begin{array}{r} 12 \\ \times 28 \\ \hline \end{array} \quad \begin{array}{r} 42 \\ \times 39 \\ \hline \end{array} \quad \begin{array}{r} 33 \\ \times 68 \\ \hline \end{array} \quad \begin{array}{r} 45 \\ \times 45 \\ \hline \end{array} \quad \begin{array}{r} 65 \\ \times 26 \\ \hline \end{array} \quad \begin{array}{r} 17 \\ \times 23 \\ \hline \end{array}$$

$$\begin{array}{r} 18 \\ \times 22 \\ \hline \end{array} \quad \begin{array}{r} 19 \\ \times 33 \\ \hline \end{array} \quad \begin{array}{r} 48 \\ \times 47 \\ \hline \end{array} \quad \begin{array}{r} 73 \\ \times 18 \\ \hline \end{array} \quad \begin{array}{r} 19 \\ \times 52 \\ \hline \end{array} \quad \begin{array}{r} 58 \\ \times 13 \\ \hline \end{array} \quad \begin{array}{r} 17 \\ \times 74 \\ \hline \end{array}$$

FIGURE 4-2: A worksheet for two-digit multiplication.

$$\begin{array}{r} 19 \\ \times 28 \\ \hline \end{array} \quad \begin{array}{r} 13 \\ \times 78 \\ \hline \end{array} \quad \begin{array}{r} 14 \\ \times 56 \\ \hline \end{array} \quad \begin{array}{r} 17 \\ \times 73 \\ \hline \end{array} \quad \begin{array}{r} 34 \\ \times 28 \\ \hline \end{array} \quad \begin{array}{r} 25 \\ \times 19 \\ \hline \end{array} \quad \begin{array}{r} 17 \\ \times 19 \\ \hline \end{array}$$

Solving three-digit multiplication problems

In the previous section, your child practiced two-digit multiplication. In this section, they perform similar operations with larger three-digit numbers that have values for hundreds, tens, and ones.

Start with this example:

```
  133
×217
```

1. Ask your child to start with the ones column, multiplying 7x133. Your child should get:

```
   22
  133
×217
  931
```

2. Your child can move to the tens column. Have them first write the 0 in the ones column of row 2:

```
   22
  133
×217
  931
    0
```

3. Your child can now multiply 1x133. They should get:

```
   22
  133
×217
  931
 1330
```

4. Your child is now ready to multiply the hundreds column. To start, have them write two 0s for the tens and ones columns of row 3:

```
   22
  133
×217
  931
 1330
   00
```

5. Your child now multiplies 2x133. They should get:

```
   22
  133
×217
  931
 1330
26600
```

6. Your child now adds up the rows for the final result:

```
   22
  133
×217
  931
 1330
26600
28861
```

Present the following problems to your child, helping them as necessary:

$$
\begin{array}{r} 213 \\ \times\,106 \end{array}
\qquad
\begin{array}{r} 107 \\ \times\,311 \end{array}
\qquad
\begin{array}{r} 156 \\ \times\,213 \end{array}
$$

They should get the following:

$$
\begin{array}{r} 1 \\ 213 \\ \times\,106 \\ \hline 1278 \\ 0 \\ 21300 \\ \hline 22578 \end{array}
\qquad
\begin{array}{r} 107 \\ \times\,311 \\ \hline 107 \\ 1070 \\ 32100 \\ \hline 33277 \end{array}
\qquad
\begin{array}{r} 11 \\ 156 \\ \times\,213 \\ \hline 468 \\ 1560 \\ 31200 \\ \hline 33228 \end{array}
$$

If they do not, take time to work through the previous steps to solve the problems with your child.

FIND ONLINE

This book's companion website at www.dummies.com/go/teachingyourkidsnew math6-8fd contains a worksheet (the first few rows of which are shown in Figure 4-3) that your child can use to practice three-digit multiplication. Download and print the worksheet. Help your child solve the first few and then ask them to solve the rest.

$$
\begin{array}{r} 217 \\ \times\,124 \end{array}
\quad
\begin{array}{r} 172 \\ \times\,228 \end{array}
\quad
\begin{array}{r} 412 \\ \times\,399 \end{array}
\quad
\begin{array}{r} 383 \\ \times\,681 \end{array}
\quad
\begin{array}{r} 145 \\ \times\,455 \end{array}
\quad
\begin{array}{r} 165 \\ \times\,226 \end{array}
\quad
\begin{array}{r} 173 \\ \times\,237 \end{array}
$$

$$
\begin{array}{r} 184 \\ \times\,226 \end{array}
\quad
\begin{array}{r} 198 \\ \times\,332 \end{array}
\quad
\begin{array}{r} 438 \\ \times\,477 \end{array}
\quad
\begin{array}{r} 713 \\ \times\,128 \end{array}
\quad
\begin{array}{r} 159 \\ \times\,552 \end{array}
\quad
\begin{array}{r} 558 \\ \times\,143 \end{array}
\quad
\begin{array}{r} 178 \\ \times\,743 \end{array}
$$

FIGURE 4-3:
A worksheet for practicing three-digit multiplication.

$$
\begin{array}{r} 193 \\ \times\,288 \end{array}
\quad
\begin{array}{r} 137 \\ \times\,785 \end{array}
\quad
\begin{array}{r} 145 \\ \times\,565 \end{array}
\quad
\begin{array}{r} 175 \\ \times\,736 \end{array}
\quad
\begin{array}{r} 347 \\ \times\,287 \end{array}
\quad
\begin{array}{r} 256 \\ \times\,196 \end{array}
\quad
\begin{array}{r} 175 \\ \times\,198 \end{array}
$$

Preparing Your Child for Success with New-Math Multiplication

New-math multi-digit multiplication uses a technique called *box multiplication*, which this section describes in detail. Consider the following problem:

$$
\begin{array}{r} 35 \\ \times\,17 \end{array}
$$

Using box multiplication, you first create a box that breaks down the problem into several smaller multiplication problems:

	30	5
20		
7		

You first multiply the box numbers, writing the results in the corresponding box locations:

	30	5
20	600	100
7	210	35

Finally, you add up the numbers in the boxes to get the result:

```
  600
  100
  210
+  35
  945
```

As you might suspect, multiplying large numbers is not something your child can easily do in their head. Instead, your child can use old-school techniques to multiply each number:

	30	5
20		
7		

```
   30      20      30      7
 × 20     × 5     × 7     × 5
    0     100     210      35
  600
  600
```

$$\begin{array}{r} 30 \\ \times 20 \\ \hline 0 \\ 600 \\ \hline 600 \end{array} \qquad \begin{array}{r} 20 \\ \times 5 \\ \hline 100 \end{array} \qquad \begin{array}{r} 30 \\ \times 7 \\ \hline 210 \end{array} \qquad \begin{array}{r} 7 \\ \times 5 \\ \hline 35 \end{array}$$

To give your child an advantage with box multiplication, you can create the following flashcards:

10 ×10	10 ×20	10 ×30	10 ×40	10 ×50	10 ×60	10 ×70	10 ×80	10 ×90	10 ×100
20 ×10	20 ×20	20 ×30	20 ×40	20 ×50	20 ×60	20 ×70	20 ×80	20 ×90	20 ×100
30 ×10	30 ×20	30 ×30	30 ×40	30 ×50	30 ×60	30 ×70	30 ×80	30 ×90	30 ×100
40 ×10	40 ×20	40 ×30	40 ×40	40 ×50	40 ×60	40 ×70	40 ×80	40 ×90	40 ×100
50 ×10	50 ×20	50 ×30	50 ×40	50 ×50	50 ×60	50 ×70	50 ×80	50 ×90	50 ×100
60 ×10	60 ×20	60 ×30	60 ×40	60 ×50	60 ×60	60 ×70	60 ×80	60 ×90	60 ×100
70 ×10	70 ×20	70 ×30	70 ×40	70 ×50	70 ×60	70 ×70	70 ×80	70 ×90	70 ×100
80 ×10	80 ×20	80 ×30	80 ×40	80 ×50	80 ×60	80 ×70	80 ×80	80 ×90	80 ×100
90 ×10	90 ×20	90 ×30	90 ×40	90 ×50	90 ×60	90 ×70	90 ×80	90 ×90	90 ×100
100 ×10	100 ×20	100 ×30	100 ×40	100 ×50	100 ×60	100 ×70	100 ×80	100 ×90	100 ×100

TIP

At first glance, these flashcards may be intimidating. However, point out to your child that they are already familiar with much of the multiplication. For example, 40x40 is just 4x4 (16), with two additional zeros (1600). Likewise, 50x50 is 5x5 with two additional zeros (2500).

Like all flashcards, practice makes perfect and practice requires time. You should plan on practicing these flashcards daily for many sessions until your child masters them.

Multiplying Using the Box Method

The box-multiplication method breaks down a multiplication problem into several easier problems. The following sections cover using the box method for both two-digit and three-digit numbers.

Using the box method to multiply two-digit numbers

Present the following two-digit multiplication problem to your child:

$$\begin{array}{r} 17 \\ \times 25 \\ \hline \end{array}$$

1. **Have your child draw a box, writing down the corresponding tens and ones as shown here:**

	10	7
20		
5		

2. **Have your child multiply 20x10 and write down the result. Your child can use old-school multiplication as shown here:**

    $$\begin{array}{r} 20 \\ \times 10 \\ \hline 0 \\ 200 \\ \hline 200 \end{array}$$

	10	7
20	200	
5		

3. **Have your child multiply 20x7, writing the result in the box. Again, your child can use old-school multiplication as necessary:**

    $$\begin{array}{r} 20 \\ \times 7 \\ \hline 140 \end{array}$$

	10	7
20	200	140
5		

4. Have your child multiply 5x10, writing the result:

	10	7
20	200	140
5	50	

5. Have your child multiply 5x7, writing the result:

	10	7
20	200	140
5	50	35

6. Have your child add up the numbers in each box for the final result. They should get:

```
  200
  140
   50
+  35
  425
```

Ask your child to use box multiplication to solve for the following:

$$\begin{array}{r} 13 \\ \times\,15 \end{array} \qquad \begin{array}{r} 22 \\ \times\,33 \end{array} \qquad \begin{array}{r} 17 \\ \times\,54 \end{array}$$

They should get:

$$\begin{array}{r} 13 \\ \times\,15 \end{array}$$

	10	3
10	100	30
5	50	15

```
  100
   30
   50
   15
  195
```

$$\begin{array}{r} 22 \\ \times\,33 \end{array}$$

	20	2
30	600	60
3	60	6

```
  600
   60
   60
    6
  726
```

$$\begin{array}{r} 17 \\ \times\,54 \end{array}$$

	10	7
50	500	350
4	40	28

```
  500
  350
   40
   28
  918
```

If they do not, take time to work through each problem with your child.

FIND ONLINE

This book's companion website at www.dummies.com/go/teachingyourkidsnew math6-8fd contains a worksheet (the first few rows of which are shown in Figure 4-4) that your child can use to solve multiplication problems using the box method. Download and print the worksheet. Help your child solve the first few and then ask them to complete the rest.

17	12	45	65
× 24	× 28	× 45	× 26

FIGURE 4-4: A worksheet for using box multiplication.

18	19	48	73
× 22	× 33	× 47	× 18

REMEMBER

Multi-digit multiplication using the box method takes practice. Have your child practice with this worksheet daily until they master it. Do not move on to division until they do.

Using the box method to multiply large numbers

In the previous section, your child used box multiplication to solve two-digit multiplication problems. You can also use the box method for three-digit numbers. The challenge, however, is that the numbers can get big!

Consider the following three-digit multiplication problem:

135
× 214

To start, you create a box to hold the corresponding hundreds, tens, and ones values:

	100	30	5
200			
10			
4			

1. **Start with 200x100 using old-school multiplication:**

$$
\begin{array}{r}
200 \\
\times\ 100 \\
\hline
0 \\
0 \\
20000 \\
\hline
20000
\end{array}
$$

	100	30	5
200	20000		
10			
4			

2. **Multiply 200x30, writing the result in the corresponding box:**

$$
\begin{array}{r}
200 \\
\times\ 30 \\
\hline
0 \\
6000 \\
\hline
6000
\end{array}
$$

3. **Multiply 200x5, writing the result:**

$$
\begin{array}{r}
200 \\
\times\ 5 \\
\hline
1000
\end{array}
$$

4. **Repeat this process for the remaining boxes. You should get:**

	100	30	5
200	20000	6000	1000
10	1000	300	50
4	400	120	20

5. Add up the boxes to get your final result:

```
  20000
   6000
   1000
   1000
    300
     50
    400
    120
 +   20
  28890
```

Working Out Multiplication Word Problems

Word problems are back! In this section, your child solves word problems that require multiplication. The problems start with single-digit multiplication and then advance. As you work each problem with your child, keep the word-problem steps in mind:

1. Determine the question being asked.

2. Determine the type of math operation required, such as multiplication.

3. Optionally, draw a picture to represent the problem.

Present the following word problem to your child:

> Sarah mows lawns to earn money. She charges $5 a lawn. This week, Sarah mowed 3 lawns. How much money did Sarah earn?

Your child should identify and solve the following expression:

$$\begin{array}{r} 5 \\ \times\ 3 \\ \hline 15 \end{array}$$

If they do not, try drawing a picture:

$$5 + 5 + 5 = 15$$
$$3 \times 5 = 15$$

Present the following word problem to your child:

> The school's math club sold tickets to a dance for $2 each. The club sold 9 tickets the first day. How much money did the club make?

Your child should identify and solve the following expression:

$$\begin{array}{r} 9 \\ \times\ 2 \\ \hline 18 \end{array}$$

Solving word problems that require two-digit multiplication

In this section, your child solves word problems that require two-digit multiplication. They can use old-school multiplication or the box method to determine the results.

Present the following word problem to your child:

> The science club started a new rock collection. Each of the club's 14 members brought in 12 rocks. How many rocks did the club collect?

Your child should identify and solve the following expression:

$$\begin{array}{r} 14 \\ \times\ 12 \\ \hline 28 \\ 140 \\ \hline 168 \end{array}$$

If they do not, try drawing a picture:

$$\begin{array}{r} 14 \\ \times\ 12 \\ \hline 28 \\ 140 \\ \hline 168 \end{array}$$

Present the following word problem to your child:

> The school had a reading contest. The book club read 15 books a day for 30 days. How many books did the club read?

Your child should identify and solve the following expression:

$$
\begin{array}{r}
15 \\
\times\,30 \\
\hline
0 \\
450 \\
\hline
450
\end{array}
$$

Solving three-digit multiplication word problems

In this section, your child solves word problems that require three-digit multiplication. The numbers may get big, but the approach to the word problem remains the same.

Present the following word problem to your child:

> The State of Texas held a band contest. They invited 150 students from 110 schools. How many students, in total, did the state invite?

Your child should identify and solve the following expression:

$$
\begin{array}{r}
150 \\
\times\,110 \\
\hline
0 \\
1500 \\
15000 \\
\hline
16500
\end{array}
$$

Present the following word problem to your child:

> A summer camp association charges $100 per camper for a week-long camp. They expect 200 campers to attend. How much money will the association receive?

Your child should identify and solve the following expression:

$$
\begin{array}{r}
100 \\
\times\;\;200 \\
\hline
0 \\
00 \\
20000 \\
\hline
20000
\end{array}
$$

IN THIS CHAPTER

» **Performing common division problems in their head**

» **Getting started with long division**

» **Resolving division problems that have a remainder**

» **Solving word problems that require division**

» **Checking their work**

Chapter **5**

Divide and Conquer

D ivision, that's the opposite of multiplication, right? You can think of division as the process of dividing something into parts.

This chapter examines division, starting with flashcards to help your child learn to solve common division operations in their head. In so doing, they build a foundation they can leverage for multi-digit division. After they master flashcards, you bring in some timed division tests, so they can show off a little.

Then you and your child examine long division, which allows them to divide larger numbers. In the process, you and your child go over remainders when division isn't always exact.

And, as you might have guessed, the chapter wraps up with word problems that require your child to apply their division skills.

So, get ready for the long and short of division.

Performing Common Division Operations in Their Head

Throughout Chapters 2 to 4, your child has mastered many addition, subtraction, and multiplication operations in their head. In this section, your child again uses flashcards, this time to master division.

Working with flashcards to master basic division operations

Using your 3x5 index cards, create the following flashcards. (Alternatively, you can purchase a pack of flashcards.)

1 ÷ 1	2 ÷ 1	3 ÷ 1	4 ÷ 1	5 ÷ 1	6 ÷ 1	7 ÷ 1	8 ÷ 1	9 ÷ 1	10 ÷ 1
11 ÷ 1	12 ÷ 1	13 ÷ 1	14 ÷ 1	15 ÷ 1	16 ÷ 1	17 ÷ 1	18 ÷ 1	19 ÷ 1	20 ÷ 1
2 ÷ 2	4 ÷ 2	6 ÷ 2	8 ÷ 2	10 ÷ 2	12 ÷ 2	14 ÷ 2	16 ÷ 2	18 ÷ 2	20 ÷ 2
3 ÷ 3	6 ÷ 3	9 ÷ 3	12 ÷ 3	15 ÷ 3	18 ÷ 3	21 ÷ 3	24 ÷ 3	27 ÷ 3	30 ÷ 3
4 ÷ 4	8 ÷ 4	12 ÷ 4	16 ÷ 4	20 ÷ 4	24 ÷ 4	28 ÷ 4	32 ÷ 4	36 ÷ 4	40 ÷ 4
5 ÷ 5	10 ÷ 5	15 ÷ 5	20 ÷ 5	25 ÷ 5	30 ÷ 5	35 ÷ 5	40 ÷ 5	45 ÷ 5	50 ÷ 5
6 ÷ 6	12 ÷ 6	18 ÷ 6	24 ÷ 6	30 ÷ 6	36 ÷ 6	42 ÷ 6	48 ÷ 6	54 ÷ 6	60 ÷ 6
7 ÷ 7	14 ÷ 7	21 ÷ 7	28 ÷ 7	35 ÷ 7	42 ÷ 7	49 ÷ 7	56 ÷ 7	63 ÷ 7	70 ÷ 7
8 ÷ 8	16 ÷ 8	24 ÷ 8	32 ÷ 8	40 ÷ 8	48 ÷ 8	56 ÷ 8	64 ÷ 8	72 ÷ 8	80 ÷ 8
9 ÷ 9	18 ÷ 9	27 ÷ 9	36 ÷ 9	45 ÷ 9	54 ÷ 9	63 ÷ 9	72 ÷ 9	81 ÷ 9	90 ÷ 9
10 ÷ 10	20 ÷ 10	30 ÷ 10	40 ÷ 10	50 ÷ 10	60 ÷ 10	70 ÷ 10	80 ÷ 10	90 ÷ 10	100 ÷ 10

If you want your child to practice the flashcards on their own, write the correct answer on the back of each card so they can quickly check their result.

These flashcards all have results that are in the range of 0 to 10. By helping your child master these flashcards, you lay the foundation they will use when they perform multi-digit division.

Mastering timed division tests

Your goal in practicing with division flashcards is for your child to quickly and accurately solve key division problems in their head. After your child has mastered the flashcards, have them practice timed worksheets similar to that shown in Figure 5-1, which I have provided on this book's companion website at www.dummies.com/go/teachingyourkidsnewmath6–8fd.

Ideally, your child should accurately complete the worksheet in 5 minutes or less. When they can do so, you can move on.

Timed tests can be stressful, so try to keep the process fun. Be positive about the problems your child gets right. If they miss several problems, that's your sign that they should spend more time with the flashcards.

Your child's ability to solve common division problems in their head will build their confidence and lay the foundation for future math success.

Performing Long Division

Long division is a division operation that breaks a large division problem into a series of smaller division operations. In this section, your child reviews long division.

Present the following division problem to your child:

$$5\overline{)75}$$

1. **Have your child try to divide 5 into 7, the first number in 75. Because 5 goes into 7 one time, write 1 above the division symbol:**

$$5\overline{)\overset{1}{75}}$$

2 ÷ 2	4 ÷ 2	12 ÷ 3	18 ÷ 6	4 ÷ 4	15 ÷ 5	9 ÷ 3
8 ÷ 2	18 ÷ 3	4 ÷ 4	42 ÷ 6	18 ÷ 9	5 ÷ 1	7 ÷ 7
16 ÷ 8	63 ÷ 7	20 ÷ 5	14 ÷ 7	24 ÷ 8	45 ÷ 9	81 ÷ 9
35 ÷ 5	36 ÷ 9	40 ÷ 10	36 ÷ 6	45 ÷ 5	64 ÷ 8	56 ÷ 8
1 ÷ 1	12 ÷ 2	15 ÷ 3	28 ÷ 7	32 ÷ 8	27 ÷ 9	24 ÷ 3
36 ÷ 9	63 ÷ 7	6 ÷ 1	60 ÷ 6	20 ÷ 4	9 ÷ 3	10 ÷ 2
10 ÷ 10	10 ÷ 5	40 ÷ 10	40 ÷ 8	36 ÷ 4	15 ÷ 5	21 ÷ 3

FIGURE 5-1:
A timed division
worksheet.

2. **Have your child multiply 5x1 and subtract that result (5) from 7, as shown here:**

$$5\overline{)\,75}^{\,1}$$
$$\underline{-5}$$
$$2$$

3. **Have your child try to divide 5 into 2. Because they cannot, bring down the next number (in this case, 5), as shown here:**

$$5\overline{)\,75}^{\,1}$$
$$\underline{-5}$$
$$25$$

4. **Have your child try to divide 5 into 25. Because 5 goes into 25 five times, have your child write a 5 above the division symbol:**

$$5\overline{)\,75}^{\,15}$$
$$\underline{-5}$$
$$25$$

5. **Have your child multiply 5x5, subtracting the result:**

$$5\overline{)\,75}^{\,15}$$
$$\underline{-5}$$
$$25$$
$$\underline{-25}$$
$$0$$

Because the result of the subtraction is 0 and there are no remaining numbers under the division symbol to bring down, you are done.

Present the following problem to your child:

$$7\overline{)\,84}$$

1. **Have your child try to divide 7 into 8. Because 7 goes into 8 one time, write a 1 above the division symbol:**

$$7\overline{)\,84}^{\,1}$$

2. **Have your child multiply 7x1 and subtract that result from 8:**

$$7\overline{)\,84}^{\,1}$$
$$\underline{-7}$$
$$1$$

3. Because your child cannot divide 7 into 1 (the result), bring down the 4:

$$7)\overline{84} \\ \begin{array}{r}1\\-7\\\hline 14\end{array}$$

4. Have your child divide 7 into 14, the result of which is 2. Write the 2 above the division symbol:

$$\begin{array}{r}12\\7)\overline{84}\\-7\\\hline 14\end{array}$$

5. Have your child multiply 7x2 and subtract the result (14):

$$\begin{array}{r}12\\7)\overline{84}\\-7\\\hline 14\\-14\\\hline 0\end{array}$$

Because the result is 0 and there are no remaining numbers under the division symbol to bring down, you are done.

Present the following division problem to your child:

$$13)\overline{169}$$

1. Have your child try to divide 13 into 1. Because you can't, have them try 13 into 16, which is 1. Write the 1 above the division symbol:

$$13)\overline{169}^{\,1}$$

2. Have your child multiply 1x13 and subtract the result:

$$\begin{array}{r}1\\13)\overline{169}\\-13\\\hline 3\end{array}$$

3. Because your child cannot divide 13 into 3, bring down the 9:

$$\begin{array}{r}1\\13)\overline{169}\\-13\\\hline 39\end{array}$$

4. In this case, 13 goes into 39 three times. Write the 3 above the division symbol:

$$
\begin{array}{r}
13 \\
13\overline{)169} \\
-13 \\
\hline
39
\end{array}
$$

5. Have your child multiply 3x13 and subtract the result (39):

$$
\begin{array}{r}
13 \\
13\overline{)169} \\
-13 \\
\hline
39 \\
-39 \\
\hline
0
\end{array}
$$

Because the result is 0 and there are no remaining numbers under the division symbol to bring down, you are done.

Ask your child to complete the following division problems, helping them as necessary:

$$
3\overline{)39} \qquad 6\overline{)96} \qquad 9\overline{)108}
$$

Your child should get:

$$
\begin{array}{r}
13 \\
3\overline{)39} \\
-3 \\
\hline
09 \\
-9 \\
\hline
0
\end{array}
\qquad
\begin{array}{r}
16 \\
6\overline{)96} \\
-6 \\
\hline
36 \\
-36 \\
\hline
0
\end{array}
\qquad
\begin{array}{r}
12 \\
9\overline{)108} \\
-9 \\
\hline
18 \\
-18 \\
\hline
0
\end{array}
$$

If your child has difficulty performing long division, you may want to consider my book, *Teaching Your Kids New Math, K–5 For Dummies*. A review of the book's contents may strengthen your child's math foundation.

FIND ONLINE

This book's companion website at www.dummies.com/go/teachingyourkidsnew math6-8fd contains a division worksheet (the first few rows of which are shown in Figure 5-2) that your child can use to practice long-division operations. Download and print the worksheet. Help your child solve the first few and then ask them to complete the rest.

$$
4\sqrt{64} \qquad\qquad 7\sqrt{84} \qquad\qquad 3\sqrt{36}
$$

FIGURE 5-2:
A long-division
worksheet.

$$
4\sqrt{52} \qquad\qquad 5\sqrt{75} \qquad\qquad 6\sqrt{66}
$$

Performing Long Division with Large Numbers

While the previous section provides a basic review of long-division operations, this section asks your child to perform the same steps but with larger numbers.

Present the following problem to your child:

$$15\overline{)3765}$$

1. **Because your child cannot divide 15 into 3, have them try 15 into 37. In this case, 15 goes into 37 two times. Write the 2 above the division symbol:**

$$15\overline{)3765}^{\,2}$$

2. **Have your child multiply 2x15, subtracting the result (30) as shown:**

```
        2
15) 3765
   -30
     7
```

3. **Because your child cannot divide 15 into 7, bring down the 6:**

```
        2
15) 3765
   -30
     76
```

4. **Have your child divide 15 into 76, which is 5. Write the 5 above the division symbol:**

```
       25
15) 3765
   -30
     76
```

5. **Have your child multiply 5x15 and subtract the result (75), as shown:**

```
       25
15) 3765
   -30
     76
   -75
     1
```

6. **Because your child cannot divide 15 into 1, bring down the 5:**

```
       25
15) 3765
   -30
     76
   -75
     15
```

7. Have your child divide 15 into 15, which is 1. Write the 1 above the division symbol:

$$
\begin{array}{r}
251 \\
15\overline{)\,3765} \\
-30 \\
\hline
76 \\
-75 \\
\hline
15
\end{array}
$$

8. Ask your child to multiply 15x1, subtracting the result, 15, as shown:

$$
\begin{array}{r}
251 \\
15\overline{)\,3765} \\
-30 \\
\hline
76 \\
-75 \\
\hline
15 \\
-15 \\
\hline
0
\end{array}
$$

Because the result is 0 and there are no remaining numbers to bring down, you are done.

Ask your child to complete the following problems:

$$12\overline{)\,1524} \qquad 23\overline{)\,8395} \qquad 44\overline{)\,11176}$$

They should get:

$$
\begin{array}{r}
129 \\
12\overline{)\,1524} \\
-12 \\
\hline
32 \\
-24 \\
\hline
84 \\
-84 \\
\hline
0
\end{array}
\qquad
\begin{array}{r}
365 \\
23\overline{)\,8395} \\
-69 \\
\hline
149 \\
-138 \\
\hline
115 \\
-115 \\
\hline
0
\end{array}
\qquad
\begin{array}{r}
254 \\
44\overline{)\,11176} \\
-88 \\
\hline
237 \\
-220 \\
\hline
176 \\
-176 \\
\hline
0
\end{array}
$$

FIND ONLINE

This book's companion website at www.dummies.com/go/teachingyourkidsnew math6–8fd contains a division worksheet (the first few rows of which are shown in Figure 5-3) that your child can use to practice long-division operations. Download and print the worksheet. Help your child solve the first few and then ask them to complete the rest.

$$13\sqrt{169} \qquad 15\sqrt{225} \qquad 17\sqrt{221}$$

FIGURE 5-3:
A long-division worksheet.

$$15\sqrt{1845} \qquad 12\sqrt{1440} \qquad 16\sqrt{5136}$$

Solving Division Problems That Have a Remainder

Unfortunately, not all division problems have an exact result where the first number divides evenly into the larger number, and so you're left with a remainder. This section provides practice on dividing with a remainder.

Present the following division problem to your child:

$$3\overline{)37}$$

1. **Have your child divide 3 into 3, which is 1. Write the 1 above the division symbol:**

$$3\overline{)37}^{1}$$

2. **Ask your child to multiply 3x1 and to subtract the result:**

$$\begin{array}{r} 1 \\ 3\overline{)\ 37} \\ \underline{-3} \\ 0 \end{array}$$

3. **Because your child cannot divide 3 into 0, bring down the 7:**

$$\begin{array}{r} 1 \\ 3\overline{)\ 37} \\ \underline{-3} \\ 07 \end{array}$$

4. **Have your child divide 3 into 7, which is 2. Write the 2 above the division symbol:**

$$\begin{array}{r} 12 \\ 3\overline{)\ 37} \\ \underline{-3} \\ 07 \end{array}$$

5. **Ask your child to multiply 3x2 and subtract the result:**

$$\begin{array}{r} 12 \\ 3\overline{)\ 37} \\ \underline{-3} \\ 07 \\ \underline{-6} \\ 1 \end{array}$$

6. Because there are no more numbers to bring down, you are done. Explain to your child that 1 is called the remainder. Have your child write the remainder in the following form:

$$
\begin{array}{r}
12\ \text{R}1 \\
3{\overline{\smash{)}\,37}} \\
\underline{-3} \\
07 \\
\underline{-6} \\
1
\end{array}
$$

Present the following division problem to your child:

$$12{\overline{\smash{)}\,149}}$$

1. Because your child cannot divide 12 into 1, try 12 into 14, which is 1. Write the 1 above the division symbol:

$$
\begin{array}{r}
1 \\
12{\overline{\smash{)}\,149}}
\end{array}
$$

2. Have your child multiply 12x1 and subtract the result, as shown:

$$
\begin{array}{r}
1 \\
12{\overline{\smash{)}\,149}} \\
\underline{-12} \\
2
\end{array}
$$

3. Because your child cannot divide 12 into 2, bring down the 9:

$$
\begin{array}{r}
1 \\
12{\overline{\smash{)}\,149}} \\
\underline{-12} \\
29
\end{array}
$$

4. Have your child divide 12 into 29, which is 2. Write the 2 above the division symbol:

$$
\begin{array}{r}
12 \\
12{\overline{\smash{)}\,149}} \\
\underline{-12} \\
29
\end{array}
$$

5. Have your child multiply 12x2 and subtract the result:

$$
\begin{array}{r}
12 \\
12{\overline{\smash{)}\,149}} \\
\underline{-12} \\
29 \\
\underline{24} \\
5
\end{array}
$$

6. Because there are no remaining numbers to bring down, you are done; 5 is the remainder. Write the result in the following form:

$$
\begin{array}{r}
12\text{R}5 \\
12\overline{)149} \\
-12 \\
\hline
29 \\
24 \\
\hline
5
\end{array}
$$

FIND ONLINE

This book's companion website at www.dummies.com/go/teachingyourkidsnew math6-8fd contains a worksheet (the first few rows of which are shown in Figure 5-4) that your child can use to solve division problems that have a remainder. Download and print the worksheet. Help your child solve the first few and then ask them to complete the rest.

$$13\sqrt{172} \qquad 15\sqrt{232} \qquad 17\sqrt{223}$$

FIGURE 5-4:
A worksheet with remainder division.

$$15\sqrt{1848} \qquad 12\sqrt{1443} \qquad 16\sqrt{5141}$$

Working Out Division Word Problems

"They're back!" Word problems, that is! In this section, your child solves word problems that require division. The problems start with simple division, and then the problems get bigger.

Solving simple division word problems

Present the following word problem to your child:

Marvin has 20 nickels he wants to share with his four friends. If he shares the nickels evenly, how many nickels will each friend get?

Your child should identify and solve the following expression:

$$4\overline{)20}$$

If they do not, try drawing the following picture:

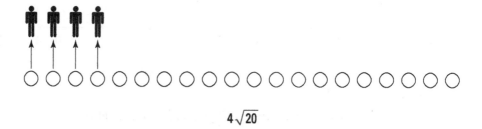

$$4\sqrt{20}$$

Present the following word problem to your child:

> Martha has a pizza that is sliced into 8 pieces. She wants to share the slices equally with her 4 friends. How many slices of pizza will each friend get?

Your child should identify and solve the following expression:

$4\overline{)8}$

If they do not, try drawing the following picture:

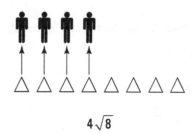

$$4\sqrt{8}$$

Solving long-division word problems

In this section, your child solves word problems that require long division. Present the following word problem to your child:

> Jan's family has 3 days to drive 1,530 miles. If they drove the same number of miles each day, how many miles should they drive each day?

Your child should identify and solve the following expression:

$3\overline{)1530}$

If they do not, draw the following picture and use it to help your child understand the previous expression:

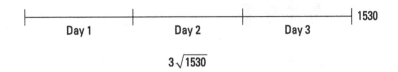

Day 1   Day 2   Day 3   1530

$$3\sqrt{1530}$$

Present the following word problem to your child:

Beth has 144 flowers that she wants to share with the 24 students in her class. How many flowers should Beth give to each student?

Your child should identify and solve the following expression:

$$24\overline{)144}$$

Present the following word problem to your child:

Jake has 108 marbles that he wants to share with his 8 friends. If Jake gives each friend the same number of marbles, how many marbles will he have left over (the remainder)?

Your child should identify the following expression:

$$8\overline{)108}$$

And here is the solution:

$$
\begin{array}{r}
13R4 \\
8\overline{)108} \\
-8 \\
\hline
28 \\
-24 \\
\hline
4
\end{array}
$$

The question is asking your child to provide the remainder.

REMEMBER

Checking Their Work

People make mistakes. When your child solves math problems, they will, too. In this section, your child learns how to check their own work.

Explain to your child that when they complete an addition problem, they can use subtraction to check their result. Ask your child to solve the following problem:

$$
\begin{array}{r}
17 \\
+14 \\
\hline
\end{array}
$$

Your child should get 31. If they do not, don't worry! Tell your child they can use subtraction to check their result:

1. **Ask your child to subtract one of the two numbers they added from their result:**

$$
\begin{array}{r}
31 \\
-17 \\
\hline
14 \\
\end{array}
$$

2. **Explain to your child that if their result was correct, then when they subtract one of the numbers they added from the result, they should get the other number.**

Present the following problem to your child:

$$
\begin{array}{r}
77 \\
-49 \\
\hline
\end{array}
$$

Your child should get 28. If they do not, don't worry.

1. **Explain to your child that when they subtract one number from another, they can use addition to check their work.**

2. **Have your child add one of the numbers they subtracted to their result:**

$$
\begin{array}{r}
28 \\
+49 \\
\hline
77 \\
\end{array}
$$

3. **Explain to your child that if the result of their addition equals the original number, they are correct.**

Present the following multiplication problem to your child:

$$
\begin{array}{r}
24 \\
\times 13 \\
\hline
\end{array}
$$

Your child should get 312. Again, if they do not, don't worry.

1. **Explain to your child that when they multiply two numbers, they can use division to check their work.**

2. Have your child divide one of the numbers they multiplied into their result:

$$
\begin{array}{r}
24 \\
13\overline{)312} \\
-26 \\
\hline
52
\end{array}
$$

3. Explain to your child that if the result of the division equals the other number they multiplied, their result is correct.

Present the following division problem to your child:

$$18\overline{)234}$$

Your child should get 13. If they do not, don't worry.

1. Explain to your child that when they divide two numbers, they can use multiplication to test their result.

2. Have your child multiply the result (13) by the original number (18):

$$
\begin{array}{r}
18 \\
\times 13 \\
\hline
54 \\
180 \\
\hline
234
\end{array}
$$

3. Explain to your child that if the result of the multiplication equals the original number, the result is correct.

As your child learned in this chapter, there are times when the result of a division operation has a remainder:

$$
\begin{array}{r}
13R1 \\
6\overline{)79}
\end{array}
$$

1. Explain to your child that when the result of a division has a remainder, they can use multiplication and addition to check the result.

2. To start, multiply the result times the original number:

$$
\begin{array}{r}
13 \\
\times 6 \\
\hline
78
\end{array}
$$

3. Add the remainder to that result:

$$
\begin{array}{r}
78 \\
+1 \\
\hline
79
\end{array}
$$

4. Explain to your child that if the result matches the original number, their result is correct.

IN THIS CHAPTER

» **Identifying fractions**

» **Adding and subtracting like fractions**

» **Multiplying fractions**

» **Dividing fractions**

» **Adding and subtracting unlike fractions**

» **Reducing fractions**

Chapter **6**

Mastering Fractions

The mere mention of the word *fractions* can create stress in adults who had hoped it was okay that they hadn't thought about adding, subtracting, multiplying, or dividing them since their own grade-school days.

Relax! Most people use fractions every day. You grab tools from your toolbox, the sizes of which are expressed as fractions; you measure things; and you can quickly recognize the bigger slice of pizza or cake.

This chapter provides a refresher on working with fractions, starting with a review of common fractions. You then go over how to add and subtract "like" fractions that have the same denominator (you remember, the same bottom number). After that, you tackle multiplying and dividing them and how to add and subtract "unlike" fractions. And finally, you look at ways to reduce fractions into their proper form — you remember, you need to reduce $\frac{2}{4}$ to $\frac{1}{2}$, $\frac{6}{8}$ to $\frac{3}{4}$, and so on.

So, get ready to dive back into fractions!

Recognizing Fractions

Ideally, your child already learned to identify fractions before sixth grade. If, for some reason, they did not, this section presents fractions to them. If your child can identify fractions, this section serves as a good review.

A *fraction* is simply part of something. Consider the following fractions:

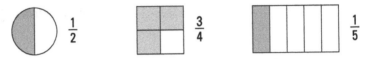

Explain to your child: "A fraction consists of two numbers, with one placed on top of the other. The top number corresponds to the number of shaded parts, and the bottom number to the total number of parts."

Ask your child to identify the following fractions:

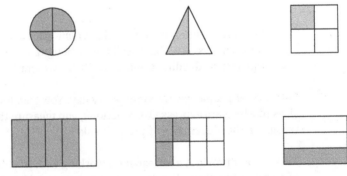

They should get:

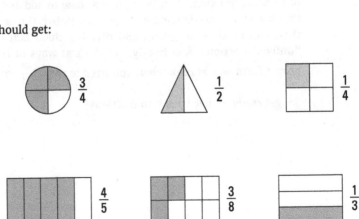

THANK THE EGYPTIANS FOR FRACTIONS

The first use of fractions dates back thousands of years to the Egyptians. I'm guessing halfway through building the pyramids someone pushing a large stone asked, "How much more do we need to build? "

The origin of the word fraction is Latin, *fractio*, which means "to break apart." So, the next time you break apart (fractio) a pizza or six-pack, be ready to impress your friends with your knowledge of fractions.

FIND ONLINE

This book's companion website at www.dummies.com/go/teachingyourkidsnew math6-8fd contains a worksheet (the first part of which is shown in Figure 6-1) that your child can use to identify various fractions. Download and print the worksheet. Help your child solve the first few and then ask them to complete the rest.

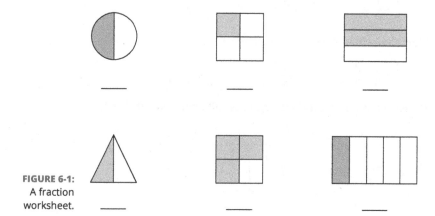

FIGURE 6-1:
A fraction
worksheet.

Adding Like Fractions

Fractions are numbers, and like other numbers, you can add, subtract, multiply, and divide them. Explain to your child that, "when two fractions have the same bottom number (which is called the *denominator*), they are 'like' fractions."

Explain to your child that they are going to practice adding like fractions.

Present the following expression to your child:

$$\frac{3}{5} + \frac{1}{5} =$$

Tell your child: "To add like fractions, you add the two top numbers while leaving the bottom number unchanged. In this case, the result becomes"

$$\frac{3}{5} + \frac{1}{5} = \frac{4}{5}$$

Visually, your child can view the operation as:

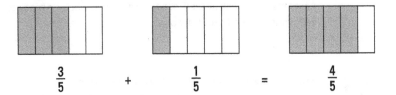

$$\frac{3}{5} \qquad + \qquad \frac{1}{5} \qquad = \qquad \frac{4}{5}$$

Ask your child to solve the following problems:

$$\frac{1}{3} + \frac{1}{3} = \qquad\qquad \frac{1}{6} + \frac{3}{6} = \qquad\qquad \frac{1}{8} + \frac{5}{8} =$$

Your child should get the following:

$$\frac{1}{3} + \frac{1}{3} = \frac{2}{3} \qquad\qquad \frac{1}{6} + \frac{3}{6} = \frac{4}{6} \qquad\qquad \frac{1}{8} + \frac{5}{8} = \frac{6}{8}$$

If they do not, present them with the following visuals:

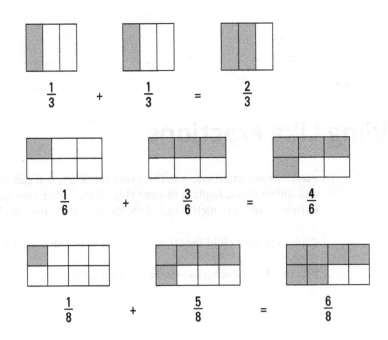

$$\frac{1}{3} \qquad + \qquad \frac{1}{3} \qquad = \qquad \frac{2}{3}$$

$$\frac{1}{6} \qquad + \qquad \frac{3}{6} \qquad = \qquad \frac{4}{6}$$

$$\frac{1}{8} \qquad + \qquad \frac{5}{8} \qquad = \qquad \frac{6}{8}$$

FIND ONLINE

This book's companion website at www.dummies.com/go/teachingyourkidsnew math6-8fd contains a worksheet (the first few rows of which are shown in Figure 6-2) that your child can use to practice adding like fractions.

$$\frac{1}{3} + \frac{1}{3} = \qquad \frac{2}{5} + \frac{1}{5} = \qquad \frac{3}{8} + \frac{4}{8} =$$

FIGURE 6-2:
A worksheet for adding like fractions.

$$\frac{1}{6} + \frac{2}{6} = \qquad \frac{1}{9} + \frac{3}{9} = \qquad \frac{1}{7} + \frac{2}{7} =$$

Download and print the worksheet. Help your child solve the first few problems and then ask them to complete the rest. If your child misses one, have them draw the corresponding fractions.

Hold off on subtracting like fractions until your child masters adding them.

TIP

Subtracting Like Fractions

Subtracting like fractions is very similar to adding them.

Tell your child: "To subtract one like fraction from another, you subtract the top numbers, leaving the bottom number unchanged."

Ask your child to solve the following expression:

$$\frac{3}{5} - \frac{1}{5} =$$

Your child should get $\frac{2}{5}$. If they do not, present the following visual:

$$\frac{3}{5} \qquad - \qquad \frac{1}{5} \qquad = \qquad \frac{2}{5}$$

Ask your child to solve the following expressions:

$$\frac{3}{5} - \frac{2}{5} = \qquad \frac{4}{7} - \frac{1}{7} = \qquad \frac{5}{6} - \frac{1}{6} =$$

Your child should get:

$$\frac{3}{5} - \frac{2}{5} = \frac{1}{5} \qquad \frac{4}{7} - \frac{1}{7} = \frac{3}{7} \qquad \frac{5}{6} - \frac{1}{6} = \frac{4}{6}$$

TIP

If your child misses a problem, try drawing the fractions that represent the corresponding expression.

**FIND
ONLINE**

This book's companion website at www.dummies.com/go/teachingyourkidsnew math6-8fd contains a worksheet (the first few rows of which are shown in Figure 6-3) that your child can use to practice subtracting like fractions. Download and print the worksheet. Help your child solve the first few and then ask them to complete the rest.

$$\frac{2}{3} - \frac{1}{3} = \qquad \frac{3}{4} - \frac{1}{4} = \qquad \frac{5}{6} - \frac{3}{6} =$$

FIGURE 6-3:
A worksheet for
subtracting like
fractions.

$$\frac{7}{8} - \frac{6}{8} = \qquad \frac{3}{9} - \frac{1}{9} = \qquad \frac{4}{10} - \frac{2}{10} =$$

Multiplying Fractions

Just as your child can add and subtract fractions, they can also multiply and divide them. In this section, your child can practice multiplying fractions.

Present the following expression to your child:

$$\frac{2}{3} \times \frac{1}{2} =$$

Explain to your child that when multiplying fractions, they multiply the two top numbers and then multiply the two bottom numbers:

$$\frac{2}{3} \xrightarrow{\times} \frac{1}{2} = \frac{2}{6}$$

Ask your child to solve the following expressions:

$$\frac{1}{2} \times \frac{1}{3} = \qquad \frac{3}{5} \times \frac{7}{8} = \qquad \frac{1}{8} \times \frac{1}{2} =$$

Your child should get:

$$\frac{1}{2} \times \frac{1}{3} = \frac{1}{6} \qquad \frac{3}{5} \times \frac{7}{8} = \frac{21}{40} \qquad \frac{1}{8} \times \frac{1}{2} = \frac{1}{16}$$

If they do not, help them perform the multiplication, first multiplying the top numbers and then multiplying the bottom numbers.

This book's companion website at www.dummies.com/go/teachingyourkidsnew math6-8fd contains a worksheet (the first few rows of which are shown in Figure 6-4) that your child can use to practice multiplying fractions.

$$\frac{1}{3} \times \frac{1}{2} = \qquad\qquad \frac{2}{3} \times \frac{1}{4} = \qquad\qquad \frac{5}{6} \times \frac{1}{3} =$$

FIGURE 6-4:
A fraction multiplication worksheet.

$$\frac{1}{2} \times \frac{1}{2} = \qquad\qquad \frac{3}{4} \times \frac{1}{3} = \qquad\qquad \frac{5}{8} \times \frac{3}{4} =$$

Download and print the worksheet. Help your child solve the first few and then ask them to complete the rest.

TIP

Focus on multiplying fractions until your child has mastered them before moving on to dividing fractions.

Dividing Fractions

Dividing fractions is very similar to multiplying — the only difference is that to divide fractions, you flip the second fraction (creating what's called a *reciprocal fraction*).

Explain to your child: "When you flip a fraction, you create a reciprocal, which has the previous fraction's top number on the bottom and bottom number on the top." For example, the reciprocal of $\frac{2}{3}$ is $\frac{3}{2}$ and the reciprocal of $\frac{4}{5}$ is $\frac{5}{4}$.

Ask your child to identify the reciprocal of the following fractions:

$$\frac{1}{4} \qquad\qquad \frac{3}{5} \qquad\qquad \frac{7}{8}$$

They should get:

$$\frac{1}{4} \quad \frac{4}{1} \qquad\qquad \frac{3}{5} \quad \frac{5}{3} \qquad\qquad \frac{7}{8} \quad \frac{8}{7}$$

If they do not, walk your child through the process of flipping a fraction by rewriting the top number on the bottom and the bottom number on the top.

Once you're confident your child has a handle on flipping fractions, present the following expression to your child:

$$\frac{1}{3} \div \frac{1}{2} =$$

1. **Have your child rewrite the expression using multiplication and the reciprocal of the second value:**

$$\frac{1}{3} \div \frac{1}{2} = \frac{1}{3} \times \frac{2}{1}$$

2. **Have your child solve the multiplication expression:**

$$\frac{1}{3} \times \frac{2}{1} = \frac{2}{3}$$

Repeat this process for the following expression:

$$\frac{1}{4} \div \frac{1}{3} =$$

1. **Rewrite the expression using multiplication and the reciprocal of the second value:**

$$\frac{1}{4} \div \frac{1}{3} = \frac{1}{4} \times \frac{3}{1}$$

2. **Have your child solve the multiplication problem.**

$$\frac{1}{4} \times \frac{3}{1} = \frac{3}{4}$$

Ask your child to solve the following expressions:

$$\frac{1}{3} \div \frac{2}{3} = \qquad\qquad \frac{1}{9} \div \frac{1}{2} = \qquad\qquad \frac{3}{4} \div \frac{2}{3} =$$

Your child should get:

$$\frac{1}{3} \div \frac{2}{3} = \frac{1}{3} \times \frac{3}{2} = \frac{3}{6}$$

$$\frac{1}{9} \div \frac{1}{2} = \frac{1}{9} \times \frac{2}{1} = \frac{2}{9}$$

$$\frac{3}{4} \div \frac{2}{3} = \frac{3}{4} \times \frac{3}{2} = \frac{9}{8}$$

If they do not, help your child use the reciprocal of the second fraction and then perform the multiplication.

FIND ONLINE

This book's companion website at www.dummies.com/go/teachingyourkidsnew math6-8fd contains a worksheet (the first few rows of which are shown in Figure 6-5) that your child can use to practice dividing fractions. Download and print the worksheet. Help your child solve the first few and then ask them to complete the rest.

$$\frac{1}{2} \div \frac{1}{4} = \qquad \frac{1}{6} \div \frac{3}{4} = \qquad \frac{1}{2} \div \frac{1}{4} =$$

FIGURE 6-5:
A worksheet for
dividing fractions.

$$\frac{3}{4} \div \frac{1}{3} = \qquad \frac{1}{8} \div \frac{7}{8} = \qquad \frac{5}{6} \div \frac{1}{3} =$$

Adding and Subtracting Unlike Fractions

Unlike fractions are fractions with different denominators. The fractions $\frac{1}{5}$ and $\frac{1}{6}$ are unlike fractions. This chapter provides a quick review of how to add and subtract unlike fractions.

Adding unlike fractions

To add (and subtract) unlike fractions, you must first convert the unlike fractions to like fractions.

Consider the following expression:

$$\frac{3}{5} + \frac{1}{3} =$$

Explain to your child: "Because the fractions have different denominators, the fractions are 'unlike.' Before you can perform the addition, you must first convert the fractions so they have the same denominator."

1. Multiply the first fraction's top and bottom number by the second fraction's denominator, which in this case is 3:

$$\frac{3}{5} \times \frac{3}{3} = \frac{9}{15}$$

2. Multiply the second fraction's top and bottom number by the first fraction's denominator, which in this case is 5:

$$\frac{1}{3} \times \frac{5}{5} = \frac{5}{15}$$

3. Now that the fractions are like, perform the addition:

$$\frac{9}{15} + \frac{5}{15} = \frac{14}{15}$$

Help your child complete the following expression:

$$\frac{1}{2} + \frac{2}{5} =$$

1. Multiply the first fraction's top and bottom number by the second fraction's denominator, which in this case is 5:

$$\frac{1}{2} \times \frac{5}{5} = \frac{5}{10}$$

2. Multiply the second fraction's top and bottom number by the first fraction's denominator, which in this case is 2:

$$\frac{2}{5} \times \frac{2}{2} = \frac{4}{10}$$

3. Now that the fractions are like, perform the addition:

$$\frac{5}{10} + \frac{4}{10} = \frac{9}{10}$$

FIND ONLINE

This book's companion website at www.dummies.com/go/teachingyourkidsnew math6-8fd contains a worksheet (the first few rows of which are shown in Figure 6-6) that your child can use to practice adding unlike fractions. Download and print the worksheet. Help your child solve the first few and then ask them to complete the rest.

$$\frac{1}{2}+\frac{1}{3}=$$ $$\frac{1}{4}+\frac{1}{2}=$$ $$\frac{3}{8}+\frac{1}{4}=$$

FIGURE 6-6:
A worksheet for
adding unlike
fractions.

$$\frac{6}{7}+\frac{1}{2}=$$ $$\frac{1}{3}+\frac{3}{4}=$$ $$\frac{1}{6}+\frac{3}{5}=$$

Subtracting unlike fractions

Explain to your child: "Just like with adding unlike fractions, to subtract unlike fractions, you must first convert the fractions to like fractions that have the same denominator."

Ask your child to consider the following expression:

$$\frac{3}{5}-\frac{1}{2}=$$

1. Multiply the first fraction's top and bottom number by the second fraction's denominator, which in this case is 2:

$$\frac{3}{5}\times\frac{2}{2}=\frac{6}{10}$$

2. Multiply the second fraction's top and bottom number by the first fraction's denominator, which in this case is 5:

$$\frac{1}{2}\times\frac{5}{5}=\frac{5}{10}$$

3. Now that the fractions are like, perform the subtraction:

$$\frac{6}{10}-\frac{5}{10}=\frac{1}{10}$$

Help your child solve the following expression:

$$\frac{3}{4}-\frac{1}{3}=$$

1. Multiply the first fraction's top and bottom number by the second fraction's denominator, which in this case is 3:

$$\frac{3}{4}\times\frac{3}{3}=\frac{9}{12}$$

2. Multiply the second fraction's top and bottom number by the first fraction's denominator, which in this case is 4:

$$\frac{1}{3}\times\frac{4}{4}=\frac{4}{12}$$

3. Now that the fractions are like, perform the subtraction:

$$\frac{9}{12} - \frac{4}{12} = \frac{5}{12}$$

FIND
ONLINE

This book's companion website at www.dummies.com/go/teachingyourkidsnew
math6–8fd contains a worksheet (the first few rows of which are shown in
Figure 6-7) that your child can use to practice subtracting unlike fractions.

$$\frac{1}{2} - \frac{1}{3} =$$ $$\frac{5}{6} - \frac{1}{4} =$$ $$\frac{7}{8} - \frac{3}{4} =$$

FIGURE 6-7:
A worksheet for
subtracting unlike
fractions.

$$\frac{5}{6} - \frac{1}{2} =$$ $$\frac{7}{8} - \frac{1}{3} =$$ $$\frac{3}{4} - \frac{2}{3} =$$

Download and print the worksheet. Help your child solve the first few and then
ask them to complete the rest.

Converting Improper Fractions

Often, when you perform math operations on fractions, you get a result for which
the fraction's top number (the numerator) is bigger than the bottom number (the
denominator), such as $\frac{5}{4}$. Such a fraction is called an *improper fraction*.

To convert an improper fraction to a proper fraction, you must convert the
fraction to a mixed number. A *mixed number* is a number that has a whole and
fractional part, such as $1\frac{1}{3}$ and $2\frac{2}{3}$.

The following expression, for example, converts $\frac{5}{4}$, an improper fraction, to its
proper form:

$$\frac{5}{4} = 1\frac{1}{4}$$

Ask your child to consider the following improper fraction:

$$\frac{5}{2}$$

1. **Explain to your child: "To convert an improper fraction to a mixed number, you divide the fraction's top number by the bottom number."**

$$2\overline{)5}$$

In this case, you get:

$$2\overline{)5}\ \ \begin{array}{l}2\,\text{R}1\\ \underline{4}\\ 1\end{array}$$

2. **Write the number to the left of the remainder, which in this case is 2, as the whole number. Then, write the remainder as a fraction using the original denominator:**

$$2\,\text{R}1\ =\ 2\frac{1}{2}$$

Consider the following improper fraction:

$$\frac{10}{3}$$

1. **Divide the fraction's top number by the bottom number:**

$$3\overline{)10}\ \ \begin{array}{l}3\,\text{R}1\\ \underline{9}\\ 1\end{array}$$

2. **Write the number to the left of the remainder, which in this case is 3, as the whole number. Then, write the remainder as a fraction using the original denominator:**

$$3\frac{1}{3}$$

FIND ONLINE

This book's companion website at www.dummies.com/go/teachingyourkidsnew math6–8fd contains a worksheet (the first few rows of which are shown in Figure 6-8) that your child can use to practice writing improper fractions as proper fractions. Download and print the worksheet. Help your child solve the first few and then ask them to complete the rest.

FIGURE 6-8:
A worksheet for converting improper fractions to their proper form.

$$\frac{5}{4} =\qquad\qquad \frac{4}{3} =\qquad\qquad \frac{7}{5} =\qquad\qquad \frac{9}{8} =$$

$$\frac{10}{8} =\qquad\qquad \frac{5}{3} =\qquad\qquad \frac{9}{7} =\qquad\qquad \frac{10}{3} =$$

Reducing Fractions to their Simplest Form

When you perform math operations with fractions, there are times when you must reduce your result into a better form—called its simplest form. For example, consider the following expression:

$$\frac{4}{10} + \frac{1}{10} = \frac{5}{10}$$

Although $\frac{5}{10}$ is correct, the fraction is not in the best form and you must further reduce it.

Explain to your child: "To reduce a fraction, you must see if you can divide any number from 2 to the numerator (the top number) into both numbers. In the case of $\frac{5}{10}$, you would check the numbers 2, 3, 4, and 5."

In this case, you can divide both the top and bottom numbers by 5:

$$\frac{5}{10} = \frac{1}{2}$$

Ask your child to consider the following fraction:

$$\frac{4}{6}$$

Say to your child: "In this case, you see if you can divide the numbers 2, 3, and 4 into the top and bottom numbers. Here, you can divide 2 into both numbers."

$$\frac{4}{6} = \frac{2}{3}$$

Ask your child to consider the fraction $\frac{4}{5}$. Say to your child: "In this case, you test whether you can divide both the fraction's top and bottom numbers by 2, 3, and 4."

Tell them: "In this case, because you cannot further divide both the top and bottom numbers, the fraction is in its fully reduced form."

Ask your child to consider the fraction:

$$\frac{20}{24}$$

Tell them: "In this case, you would try the numbers, 2, 3, 4, 20. Because you can divide both the top and bottom numbers by 2, you can reduce the fraction as follows."

$$\frac{20}{24} = \frac{10}{12}$$

Explain to your child: "Now you must check whether you can further reduce the fraction. To do so, you try the numbers 2, 3, 4, 10. In this case, you can again divide both the top and bottom numbers by 2."

$$\frac{10}{12} = \frac{5}{6}$$

Tell them: "You must again check if you can further reduce the fraction. To do so, you try the numbers 2, 3, 4, and 5. Given that you cannot further divide the fraction, it is fully reduced."

FIND ONLINE

This book's companion website at www.dummies.com/go/teachingyourkidsnew math6-8fd contains a worksheet (the first few rows of which are shown in Figure 6-9) that your child can use to practice reducing fractions.

$$\frac{6}{8} = \qquad \frac{2}{4} = \qquad \frac{2}{6} = \qquad \frac{4}{8} =$$

FIGURE 6-9: A worksheet for reducing fractions.

$$\frac{5}{10} = \qquad \frac{5}{15} = \qquad \frac{4}{6} = \qquad \frac{8}{10} =$$

Download and print the worksheet. Help your child solve the first few and then ask them to complete the rest.

Working Out Word Problems with Fractions

Word problems just wouldn't be complete if they did not include fractions. In this section, your child completes word problems that require fractions.

Present the following word problem to your child:

Tim and Linda each ate $\frac{1}{3}$ of a pizza. How much of the pizza did they eat?

Your child should identify the following expression:

$$\frac{1}{3} + \frac{1}{3} = \frac{2}{3}$$

If they do not, draw the following picture:

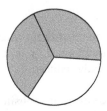

Present the following word problem to your child:

Laura read $\frac{1}{2}$ of a book and Shauna read $\frac{2}{3}$ of a book. How much did the two read in total?

Your child should identify the following expression:

$$\frac{1}{2} + \frac{2}{3} =$$

However, because the fractions are not like fractions, your child must convert them:

$$\frac{1}{2} \times \frac{3}{3} = \frac{3}{6} \qquad\qquad \frac{2}{3} \times \frac{2}{2} = \frac{4}{6}$$

$$\frac{3}{6} + \frac{4}{6} = \frac{7}{6}$$

The result is an improper fraction, which your child should reduce:

$$6)\overline{7} \quad \begin{array}{r} 1\,R1 \\ \hline \end{array} \qquad 1\frac{1}{6}$$
$$\underline{6}$$
$$1$$

Present the following word problem to your child:

At the start of Bill's trip, his gas tank was $\frac{3}{4}$ full. After the trip, the gas tank was $\frac{1}{4}$ full. How much of Bill's gas tank did he use?

Your child should identify and solve the following expression:

$$\frac{3}{4} - \frac{1}{4} = \frac{2}{4}$$

However, the result, $\frac{2}{4}$, is not fully reduced. To do so, your child should test whether they can divide both the top and bottom numbers by 2, which they can:

$$\frac{2}{4} = \frac{1}{2}$$

IN THIS CHAPTER

» **Understanding the point of decimals**

» **Addition, subtraction, multiplication, and division with decimals**

» **Repeating decimal numbers**

» **Converting fractions to decimals**

» **Dividing two decimal numbers**

» **Word problems with decimal numbers**

Chapter **7**

The Point of Decimals

I n the previous chapters, your child has reviewed addition, subtraction, multiplication, and division problems using integer numbers and fractions. In this chapter, they review those operations with decimal numbers — you know, numbers with a decimal point, such 1.1, 2.2, and 3.14. The math operations themselves don't change. The challenge is in knowing where to put the decimal point in the result.

Time to jump in and revisit the point of decimals.

Making Sense of Decimal Numbers

The first step to working with decimal numbers is being able to recognize and compare decimal values.

Say to your child: "A decimal number is a number with a decimal point, such as 1.1, 2.31, and 5.123. You use decimal numbers for a wide range of purposes, such as counting dollars and cents, measuring temperatures, and much more."

Explain to your child: "When you work with numbers, you have learned that each digit has a place, such as ones, tens, hundreds, and thousands. In a similar way, the numbers to the right of a decimal point have a place which starts with tenths, hundredths, and thousandths."

Ask your child to consider the number 5.3. Say to your child: "You can think of the number 5.3 as a fraction with the value $5\frac{3}{10}$. Likewise, you can think of 4.2 as $4\frac{2}{10}$. In a similar way, the number 2.55 is equal to $2\frac{55}{100}$, and 2.557 is equal to $2\frac{557}{1000}$."

Explain to your child that the following numbers are equal:

5.3   5.30   5.300

However, the following numbers are not equal:

5.1   5.15   5.152

Tell your child: "In this case, 5.1 is less than 5.15 and 5.15 is less than 5.152."

Ask your child to use the symbols <, =, and > to compare the following decimal numbers:

1.2	1.3	1.9	1.90	1.8	1.810
5.6	5.600	5.1	5.11	5.5	6.1
25.7	25.699	25.3	25.301	25.8	25.88

They should get the following:

1.2 < 1.3	1.9 = 1.90	1.8 < 1.810
5.6 = 5.600	5.1 < 5.11	5.5 < 6.1
25.7 > 25.699	25.3 < 25.301	25.8 < 25.88

TIP

If your child is struggling with this concept, have them represent each value as a like fraction (same denominator) and compare them.

FIND ONLINE

This book's companion website at www.dummies.com/go/teachingyourkidsnew math6–8fd contains a worksheet (the first few rows of which are shown in Figure 7-1) that your child can use to practice comparing decimal values. Download and print the worksheet. Help your child solve the first few and then ask them to complete the rest.

FIGURE 7-1:
A worksheet for comparing decimal numbers.

1.0 __ 1.1	2.5 __ 3.7	8.11 __ 8.12
3.6 __ 3.62	2.51 __ 3.42	0.2 __ 1.0
1.1 __ 1.3	3.7 __ 3.75	8.00 __ 8.01

Adding Decimal Numbers

Once you're confident that your child knows how to recognize and compare decimal numbers, you can work on adding them.

REMEMBER

The challenge of decimals is working with the two numbers to the right of the decimal point and then carrying appropriately. The number of digits to the left is just addition.

Present the following expression to your child:

```
  5.1
+3.2
```

Explain to your child: "To add two decimal values, you align the decimal point and then add the values in each place, carrying as necessary."

In this case, the numbers add up as follows:

```
  5.1
+3.2
─────
  8.3
```

Present the following expression to your child:

```
 25.3
 +8.2
```

Ask your child to solve the problem. They should get:

```
 25.3
 +8.2
─────
 33.5
```

If they do not, help your child by first adding the decimal values:

```
 25.3
+8.2
─────
   .5
```

Next, add the ones digits:

```
 1
 25.3
+8.2
─────
  3.5
```

Finally, add the tens digits:

```
 1
 25.3
+8.2
─────
 33.5
```

Explain to your child: "Depending on the numbers you are adding, there will be times when the numbers do not have the same number of digits to the right of the decimal point."

Ask your child to consider the following expression:

```
  3.1
+2.15
```

Explain to them: "When the two numbers do not have the same number of digits to the right of the decimal point, you can extend a number as necessary by adding zeros. For example, in the case of 3.1, you can use 3.10, which is an equal value."

```
 3.10
+2.15
```

Ask your child to solve the expression. They should get:

```
 3.10
+2.15
─────
 5.25
```

Ask your child to solve the following expressions:

```
 1.5        7.70        3.23
+2.3       +8.50      +1.577
```

They should get:

```
            1            1
 1.5        7.70        3.230
+2.3       +8.50      +1.577
────       ──────      ──────
 3.8       16.20        4.807
```

Take time to work on the addition operations with your child if you see that they're struggling to get the correct answer on their own.

FIND ONLINE

This book's companion website at www.dummies.com/go/teachingyourkidsnew math6–8fd contains a worksheet (the first few rows of which are shown in Figure 7-2) that your child can use to practice adding decimal numbers. Download and print the worksheet. Help your child solve the first few and then ask them to complete the rest.

1.1	2.0	3.4	5.6	7.3
+ 1.2	+ 1.3	+ 2.0	+ 2.0	+ 1.6

FIGURE 7-2: A worksheet for adding decimal numbers.

0.1	1.0	3.3	4.6	3.6
+ 0.2	+ 1.5	+ 2.15	+ 5.115	+ 4.70

Subtracting Decimal Numbers

In the previous section, your child practiced adding decimal numbers. In this section, they will subtract them.

Present the following problem to your child:

5.3
−2.1

Tell your child: "You'll find that subtracting decimal numbers is very similar to adding them. You align the decimal points and then subtract the numbers in each place, borrowing as necessary."

In this case, the previous expression becomes:

5.3
−2.1

3.2

Say to your child: "You have learned that when two numbers do not have the same number of digits to the right of the decimal point, you can extend one of the values as necessary, using zeros. For example, consider the following expressions."

3.1   4.11
−2.85   −2.5

By extending a value with zero, the expressions become:

$$\begin{array}{r} 3.10 \\ -2.85 \\ \hline \end{array} \qquad \begin{array}{r} 4.11 \\ -2.50 \\ \hline \end{array}$$

Ask your child to solve these two expressions. They should get:

$$\begin{array}{r} \overset{}{\cancel{3}}.\cancel{1}{}^{1}0 \\ -2.8\ 5 \\ \hline 0.2\ 5 \end{array} \qquad \begin{array}{r} \overset{3}{\cancel{4}}.{}^{1}11 \\ -2.\ 50 \\ \hline 1.\ 61 \end{array}$$

If your child doesn't get the right answer, it's time to review the operation with them.

FIND ONLINE

This book's companion website at www.dummies.com/go/teachingyourkidsnew math6-8fd contains a worksheet (the first few rows of which are shown in Figure 7-3) that your child can use to practice subtracting decimal numbers. Download and print the worksheet. Help your child solve the first few and then ask them to complete the rest.

$$\begin{array}{r} 2.2 \\ -1.1 \\ \hline \end{array} \qquad \begin{array}{r} 3.3 \\ -1.7 \\ \hline \end{array} \qquad \begin{array}{r} 5.5 \\ -2.2 \\ \hline \end{array} \qquad \begin{array}{r} 8.1 \\ -6.7 \\ \hline \end{array} \qquad \begin{array}{r} 9.9 \\ -6.6 \\ \hline \end{array}$$

FIGURE 7-3:
A worksheet for subtracting decimal numbers.

$$\begin{array}{r} 2.20 \\ -1.19 \\ \hline \end{array} \qquad \begin{array}{r} 3.30 \\ -2.65 \\ \hline \end{array} \qquad \begin{array}{r} 2.34 \\ -1.23 \\ \hline \end{array} \qquad \begin{array}{r} 5.5 \\ -2.25 \\ \hline \end{array} \qquad \begin{array}{r} 8.3 \\ -7.99 \\ \hline \end{array}$$

Multiplying Decimal Numbers

The multiplication process for decimal numbers is slightly different from the process for addition and subtraction. Rather than lining up the decimal points, you line up the numbers.

Present the following problem to your child:

$$\begin{array}{r} 3.1 \\ \times 2.2 \\ \hline \end{array}$$

1. Explain to your child: "When you multiply decimal numbers, you can begin by ignoring the decimal points and simply perform the multiplication."

$$
\begin{array}{r}
3.1 \\
\times\ 2.2 \\
\hline
62 \\
620 \\
\hline
682 \\
\end{array}
$$

2. Say to your child: "After you complete the multiplication, you then count the number of digits in the two numbers that appear to the right of the decimal point, which in this case is 2."

3. Say to them: "Using that count, start at the rightmost digit in your result and then move over — in this case, 2 places — and write the decimal point."

6.82

Ask your child to repeat the process to solve the following expression:

$$
\begin{array}{r}
4.3 \\
\times\ 1.7 \\
\end{array}
$$

1. Have your child perform the multiplication, ignoring the decimal points:

$$
\begin{array}{r}
4.3 \\
\times\ 1.7 \\
\hline
301 \\
430 \\
\hline
731 \\
\end{array}
$$

2. Ask your child to count the number of digits that appear to the right of the decimal point, which is 2.

3. Have your child start with the rightmost digit in the result and move to the left two locations. Write the decimal point.

7.31

Explain to your child: "When you multiply two decimal numbers, there may be times when the numbers do not have the same number of digits to the right of the decimal point." The following problem, for example, multiplies 3.25, which has two digits to the right of the decimal point and 2.1, which only has one:

$$
\begin{array}{r}
3.25 \\
\times\ 2.1 \\
\end{array}
$$

1. Say to your child: "When you multiply decimal numbers, you do not need to extend a number with zeros. Instead, you just perform the multiplication."

```
   3.25
 × 2.1
   325
  6500
  6825
```

2. Have your child count the number of digits in both numbers that appear to the right of the decimal point, which in this case is 3.

3. Have your child start with the rightmost digit in the result and move over the number of digit positions they counted in Step 2. Have them place the decimal point:

$$6.825$$

Ask your child to solve the following problems:

```
    3.3        0.7        1.35
  × 2.2      × 1.2      × 2.1
```

They should get:

```
    3.3        0.7        1.35
  × 2.2      × 1.2      × 2.1
    66         14         135
   660         70        2700
   7.26       0.84       2.835
```

If your child misses a problem, take a minute to review their math and then remind them to count digits to the right of the decimal point so they know where to put the decimal in their result.

FIND ONLINE

This book's companion website at www.dummies.com/go/teachingyourkidsnew math6-8fd contains a worksheet (the first few rows of which are shown in Figure 7-4) that your child can use to perform multiplication problems with decimals. Download and print the worksheet. Help your child solve the first few and then ask them to complete the rest.

```
   1.1        1.0        2.2        8.1        5.12
 × 1.2      × 1.3      × 1.5      × 1.0      × 1.1
```

FIGURE 7-4:
A multiplication worksheet for decimals.

```
   1.10       2.3        1.63       1.75       3.67
 × 2.2      × 4.15     × 2.0      × 2.25     × 1.23
```

TIP

Multiplying decimal numbers is a difficult process. Your child may need to practice this worksheet several times before they master the process. Do not move on to division until they do.

Dividing Decimal Numbers

In the previous section, your child practiced multiplying decimal numbers. In this section, they practice dividing them. The division process doesn't change — the challenge is getting the decimal point in the correct location.

Say to your child: "You have learned to multiply numbers with a decimal. Now you will learn to divide them."

Ask your child to consider the following problem:

$$5\overline{)35.2}$$

1. **Have your child write a decimal point above the division symbol, above the current decimal point:**

    $$5\overline{)3\overset{\cdot}{5}.2}$$

2. **In this case, you cannot divide 5 into 3, so have your child try 35. They should write 7:**

    $$\overset{7.}{5\overline{)35.2}}$$

3. **Have your child do the multiplication and subtraction:**

    $$\begin{array}{r} 7. \\ 5\overline{)\ 35.2} \\ -35 \\ \hline 0 \end{array}$$

4. **Explain to your child that when you divide decimal numbers, you don't have a remainder. Instead, you keep adding zeros to the numerator until you reach a result. Have your child bring down the 2. Because you cannot divide 5 into 2, have your child add another 0 to the numerator and bring it down:**

    $$\begin{array}{r} 7. \\ 5\overline{)\ 35.20} \\ -35 \\ \hline 020 \end{array}$$

5. Have your child divide 5 into 20, which is 4, writing the result:

```
      7.04
5) 35.20
  -35
    020
   - 20
      0
```

6. Because the result is zero and there are no more numbers to bring down, you are done.

Present the following division problem to your child:

```
6)97.5
```

1. Have your child write a decimal point above the division symbol, directly above the current decimal point:

```
  . .
6)97.5
```

2. In this case, your child can divide 6 into 9. They should write 1:

```
  1 .
6)97.5
```

3. Have your child do the multiplication and subtraction:

```
   1 .
6) 97.5
  -6
   37
```

4. Have your child divide 6 into 37. They should get 6:

```
   16.
6) 97.5
  -6
   37
  -36
   15
```

5. Have your child divide 6 into 15, which is 2:

```
   16.2
6) 97.5
  -6
   37
  -36
   15
  - 12
     3
```

6. Have your child add and bring down a zero:

$$
\begin{array}{r}
16.2 \\
6\overline{\smash{)}\,97.50} \\
\underline{-6} \\
37 \\
\underline{-36} \\
15 \\
\underline{-12} \\
30
\end{array}
$$

7. Have your child divide 6 into 30, which is 5:

$$
\begin{array}{r}
16.25 \\
6\overline{\smash{)}\,97.50} \\
\underline{-6} \\
37 \\
\underline{-36} \\
15 \\
\underline{-12} \\
30 \\
\underline{-30} \\
0
\end{array}
$$

8. Because the result is zero and there are no more numbers to bring down, you are done.

Present the following problem to your child:

$$14\overline{\smash{)}\,1901.2}$$

Although the numbers are larger, the process remains the same:

$$
\begin{array}{r}
135.8 \\
14\overline{\smash{)}\,1901.2} \\
\underline{-14} \\
50 \\
\underline{-42} \\
81 \\
\underline{-70} \\
112 \\
\underline{-112} \\
0
\end{array}
$$

Ask your child to solve the following problems:

$$12\overline{\smash{)}\,162.0} \qquad\qquad 15\overline{\smash{)}\,262.5}$$

They should get:

$$\begin{array}{r} 13.5 \\ 12\overline{)\ 162.0} \\ \underline{-12} \\ 42 \\ \underline{-36} \\ 60 \\ \underline{-60} \\ 0 \end{array} \qquad \begin{array}{r} 17.5 \\ 15\overline{)\ 262.5} \\ \underline{-15} \\ 112 \\ \underline{-105} \\ 75 \\ \underline{-75} \\ 0 \end{array}$$

If they do not, review the solution with them one step at a time, correcting any errors.

This book's companion website at www.dummies.com/go/teachingyourkidsnew math6-8fd contains a worksheet (the first few rows of which are shown in Figure 7-5) that your child can use to practice dividing decimal numbers. Download and print the worksheet. Help your child with the first few and then ask them to complete the rest.

$$5\sqrt{10.5} \qquad\qquad 7\sqrt{14.7}$$

FIGURE 7-5:
A division
worksheet for
decimals.

$$8\sqrt{800.24} \qquad\qquad 9\sqrt{1111.5}$$

Repeating Decimal Numbers

When you work with decimal numbers, there are times when you cannot exactly represent a value, and the values to the right of the decimal point repeat. For example, the fraction $\frac{1}{3}$ as a decimal is 0.33333 with the 3s repeating forever. This section provides some practice with repeating decimals.

Say to your child: "You have learned that when you divide decimal numbers, you don't get a remainder — meaning, you continue to bring down zeros until you get a result."

Say to them: "Sometimes, however, the process won't end, and instead, the numbers to the right of the decimal point repeat themselves."

Ask your child to consider the following expression:

$$3\overline{)100.0}$$

1. **Your child cannot divide 3 into 1, so try 10:**

$$
\begin{array}{r}
3 \\
3\overline{)100.0} \\
\underline{-9} \\
10
\end{array}
$$

2. **Again, have them divide 3 into 10:**

$$
\begin{array}{r}
33. \\
3\overline{)100.0} \\
\underline{-9} \\
10 \\
\underline{-9} \\
1
\end{array}
$$

3. **When they bring down the zero, they again divide 3 into 10:**

$$
\begin{array}{r}
33.3 \\
3\overline{)100.0} \\
\underline{-9} \\
10 \\
\underline{-9} \\
10 \\
\underline{-9} \\
1
\end{array}
$$

4. **Again, when they bring down the zero, they must divide 3 into 10.**

5. **Ask your child: "Can you see the numbers starting to repeat?"**

6. **Say to them: "In this case, no matter how many times you bring down a zero, the process repeats. To indicate the repeating decimal, you draw a bar above your last digit."**

$33.\overline{3}$

Ask your child to consider the following expression:

$$6\overline{)22.0}$$

1. **Your child cannot divide 6 into 2, so try 22:**

$$
\begin{array}{r}
3. \\
6\overline{)\,22.0} \\
\underline{-18} \\
40
\end{array}
$$

2. Have them divide 6 into 40, which is 6:

$$
\begin{array}{r}
3.6 \\
6\overline{\smash{)}22.0} \\
-18 \\
\hline
40 \\
-36 \\
\hline
4
\end{array}
$$

3. Again, they divide 6 into 40:

$$
\begin{array}{r}
3.66 \\
6\overline{\smash{)}22.00} \\
-18 \\
\hline
40 \\
-36 \\
\hline
40 \\
-36 \\
\hline
4
\end{array}
$$

4. And again:

$$
\begin{array}{r}
3.666 \\
6\overline{\smash{)}22.000} \\
-18 \\
\hline
40 \\
-36 \\
\hline
40 \\
-36 \\
\hline
40 \\
-36 \\
\hline
4
\end{array}
$$

5. Say to your child: "In this case, no matter how many zeros you bring down, the process simply repeats. You can represent your solution as follows."

$3.66\overline{6}$ or $3.\overline{6}$

Converting Fractions to Decimals

Every fraction has an equivalent decimal value: $\frac{3}{5}$ is 0.6, $\frac{3}{4}$ is 0.75, and so on. In this section, you and your child work on converting fractions to decimals.

Say to your child: "As you perform math operations, there will be times when you must convert a fraction to its equivalent decimal value. Every fraction has an equivalent decimal value: $\frac{1}{2}$ is 0.5, $\frac{1}{3}$ is 0.33, and $\frac{1}{4}$ is 0.25."

Explain to your child: "To convert a fraction to a decimal, you divide the bottom number into the top number. Consider the following."

$$\frac{1}{4}$$

1. Say to your child: "In this case, you divide 4 into 1."

$$4\overline{)1}$$

2. Explain to them: "Because you cannot divide 4 into 1, you add a 0 after the decimal and write a 0 above the division symbol."

$$4\overline{)1.0} \quad 0.$$

3. Have your child perform the division. They should get:

$$
\begin{array}{r}
0.25 \\
4\overline{)1.00} \\
\underline{-8} \\
20 \\
\underline{-20} \\
0
\end{array}
$$

Ask your child to consider the following example:

$$\frac{3}{4}$$

1. Say to your child: "In this case, you divide 4 into 3."

$$4\overline{)3}$$

2. Say to them: "Because you cannot divide 4 into 3, add a 0 after the decimal and write a 0 above the division symbol as follows."

$$4\overline{)3.0} \quad 0.$$

3. Have your child perform the division. They should get:

$$
\begin{array}{r}
0.75 \\
4\overline{)3.00} \\
\underline{-28} \\
20 \\
\underline{-20} \\
0
\end{array}
$$

Ask your child to convert the following fractions to their decimal equivalent values:

$$\frac{1}{2} \qquad\qquad \frac{3}{5}$$

They should get:

$$2\overline{)\begin{array}{c}0.5 \\ 1.0 \\ \underline{-10} \\ 0\end{array}}\qquad 5\overline{)\begin{array}{c}0.6 \\ 3.0 \\ \underline{-30} \\ 0\end{array}}$$

Dividing a Decimal Value into Another

While the previous section covers dividing whole numbers into a decimal value, this section provides practice in dividing a decimal value into another:

$$4.2\overline{)16.8}$$

Ask your child to consider the following expression:

$$1.3\overline{)3.9}$$

1. Explain to your child: "In this case, both numbers have decimal points. To start, you want to make the number you are dividing by a whole number. To do so, you move the decimal point to the right one location in both values."

$$1.2 \rightarrow 12 \quad 3.9 \rightarrow 39$$

$$12\overline{)39}$$

2. Now you can perform the division:

$$12\overline{)\begin{array}{c}3.25 \\ 39.00 \\ \underline{-36} \\ 30 \\ \underline{-24} \\ 60 \\ \underline{-60} \\ 0\end{array}}$$

Ask your child to consider the following expression:

$$1.25\overline{)5}$$

1. Say to your child: "To make 1.25 a whole number, you must move the decimal point in both values 2 positions to the right."

1.25 becomes 125   5.00 becomes 500

$$125\overline{)500}$$

2. Now they can perform the division. They should get:

$$125\overline{)500} \\ \underline{-500} \\ 0$$

with quotient 4.

FIND ONLINE

This book's companion website at www.dummies.com/go/teachingyourkidsnew math6–8fd contains a worksheet (the first few rows of which are shown in Figure 7-6) that your child can use to practice dividing decimal numbers. Download and print the worksheet. Help your child with the first few and then ask them to complete the rest.

$$3.3\sqrt{12.2} \qquad 1.6\sqrt{5.6} \qquad 2.2\sqrt{8.25}$$

$$7.7\sqrt{63.14} \qquad 5.5\sqrt{36.3} \qquad 9.1\sqrt{33.67}$$

FIGURE 7-6:
A worksheet for decimals.

Working through Word Problems with Decimal Numbers

A chapter would just not seem complete without word problems. In this section, your child gets to solve word problems that use decimal numbers (yay!).

Present the following word problem to your child:

> Tabitha went to the grocery store and bought candy for $1.20 and a soda for $1.50. How much did Tabitha spend at the store?

Your child should identify and solve the following problem:

$$1.20 \\ \underline{+1.50}$$

Present the following word problem to your child:

> Tim has $10.50 and spends $8.95 on a movie. How much money does Tim have left?

Your child should identify and solve the following problem:

10.50
−8.95

Present the following word problem to your child:

Jill's family went to the beach. Along the way, they stopped to buy 25 gallons of gas that cost $5.50 a gallon. How much did Jill's family pay for gas?

Your child should identify and solve the following problem:

25
×5.50

Present the following word problem to your child:

Mary visits a farm and buys 12 eggs for $3.60. How much does Mary pay for each egg?

Your child should identify and solve the following problem:

12)3.60

IN THIS CHAPTER

» **Defining the term** *factor*

» **Calculating a number's factors**

» **Identifying prime numbers**

» **Finding the greatest common factor**

» **Understanding multiples, particularly the least common multiple**

Chapter **8**

Prime Time Factors

A factor set is a set of two numbers that multiply to produce a result. For example, the factor sets of the number 4 are 1×4 and 2×2. This chapter examines factors.

Some numbers, such as 36, have many factors: 1×36, 2×18, 3×12, 4×9, and 6×6. The "prime" in this chapter's title (it's okay, you can look at the title again) refers to prime numbers for which the only factor is 1 times the number. The numbers 2, 3, 5, and 7, for example, are prime — you can only multiply 1 times the number to get the result. This chapter discusses prime numbers, which are key to data-encryption algorithms, which many of the computer programs you use today keep others from reading your secrets.

This chapter also examines the greatest common denominator for two values as well as the least common multiples, the latter of which is also used in data encryption.

So, get ready for prime time and jump in.

Understanding Factors

A *factor set* is a fancy term for two numbers that multiply together to produce a result. For example, the factor sets for the number 6 are:

1×6 2×3

Likewise, the factor sets for the number 12 are:

1×12 2×6 3×4

Say to your child: "A factor is two numbers that multiply to produce a result. Most of us use factors everyday to exchange money. For example, you might exchange four quarters for a dollar bill or I might ask someone to break a twenty-dollar bill into four five-dollar bills. In addition, many of the software programs we run use data encryption to keep our secrets safe, make use of factors — particularly, prime factors."

Present this number to your child:

24

1. Say to your child: "The factors for 24 include all the numbers that multiply to 24."

2. Ask your child to list the pairs they can multiply to equal 24. They should get:
   1×24 2×12 3×8 4×6

3. Ask your child to list the factors for the number 36.

4. They should get:
   1×36 2×18 3×12 6×6

If they do not, you may want to spend more time practicing the multiplication flashcards before you continue.

Knowing when to stop when factoring numbers

In the previous section, your child listed the factors for the numbers 24 and 36. To do so, your child listed all the pairs that multiply to equal the specific value.

Your child should start to factor a number with 1 times the number. Ask your child to consider the number 49.

1. **To list the factors, your child should first start with 1x49.**

2. **Next, your child should try the numbers 2, 3, 4, in order, to see if they factor into 49.**

3. **Your child should identify 7x7 as a factor for 49. Because the next number to consider is 8 and 8 is bigger than the previous second number (7), your child can stop.**

Ask your child to factor the number 64.

1. **To list the factors, your child should start with 1x64.**

2. **Next, your child should try the numbers 2, 3, 4, in order, to see if they factor into 64.**

3. **In this case, your child should identify:**
   1×64 2×32 4×16 8×8

4. **Because the next number to consider is 9 and 9 is bigger than the previous second number (8), your child can stop.**

**FIND
ONLINE**

This book's companion website at www.dummies.com/go/teachingyourkidsnew math6-8fd contains a worksheet (the first few rows of which are shown in Figure 8-1) that your child can use to factor numbers. Download and print the worksheet. Help your child solve the first few and then ask them to complete the rest.

1	2	3	4	5	6	7	8	9	10
11	12	13	14	15	16	17	18	19	20
21	22	23	24	25	26	27	28	29	30

FIGURE 8-1:
A factor
worksheet.

Understanding Prime Numbers

Every number always has the factor of 1 times the number. For some values, called *prime numbers*, the only factor is 1 times the number. In this section, your child practices working with prime numbers.

Say to your child: "A factor consists of two numbers that multiply to a specific result. Every number has the factor of 1 times the number."

Ask your child to consider the factors of the numbers 1 through 5:

1	2	3	4	5
1×1	1×2	1×3	$1 \times 4, 2 \times 2$	1×5

Say to them: "As you can see, every number has the factor of 1 times itself. In some cases, 1 is the only factor. You call such numbers *prime numbers*."

Ask your child to factor the following numbers and then say which numbers are prime:

15	16	17	18	19	20

They should get:

15	1×15	3×5	
16	1×16	2×8	2×8
17	1×17		
18	1×18	2×9	3×6
19	1×19		
20	1×20	2×10	4×5

In this case, your child should identify the numbers 17 and 19 as prime. Each of the other numbers has factors beyond 1 times the number.

ONE IS NOT A PRIME NUMBER

As you know, a prime number only has the factors of 1 times itself. If you consider the number 1, its only factor is 1x1. However, 1, by definition, is not considered prime. When the Greeks invented prime numbers, they wanted the number 1 to be treated as special.

Creating Factor Trees

A *factor tree* is a diagram that shows the prime numbers you can multiply to produce a result. This section allows your child to practice creating factor trees.

Say to your child: "You have learned that a prime number is special in that its only factors are 1 times the value. It turns out that for any number, there is a set of prime numbers that multiply to produce that number. To determine those prime numbers, you can use a factor tree."

Consider the value 16. To create a factor tree, you identify two factors for 16, as shown here:

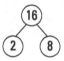

When a branch of the tree contains a prime number (such as 2 in this case), you can stop factoring. For a non-prime branch, you continue to factor:

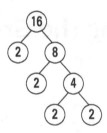

Say to your child: "In this case, if you multiply the prime numbers, the result will be 16."

$$2 \times 2 \times 2 \times 2 = 16$$

Ask your child to create a factor tree for the number 20.

They should get the following prime numbers:

Say to them: "Again, if you multiply the prime numbers, you get 20."

$$2 \times 2 \times 5 = 20$$

FIND ONLINE

This book's companion website at www.dummies.com/go/teachingyourkidsnew math6-8td contains a worksheet (the first few rows of which are shown in Figure 8-2) that your child can use to create factor trees for numbers. Download and print the worksheet. Help your child solve the first few. Then, ask them to solve the rest.

FIGURE 8-2:
A factor tree
worksheet.

8   16   24   32

Understanding the Greatest Common Factor

There may be times when your child must examine the factors for two values to determine the greatest (largest) factor the two values have in common. The greatest common factor, for example, is often used to reduce fractions.

Consider the numbers 15 and 20:

         15   20
    1×15 3×5 1×20 2×10 4×5

In this case, the greatest common factor for the numbers 15 and 20 is 5.

Say to your child: "You have learned to calculate a number's factors. As you perform various math operations, such as reducing fractions, you want to know the greatest common factor between two values."

Ask your child to consider the factors for the numbers 36 and 48:

36
1×36 2×18 3×12 4×9 6×6

48
1×48 2×24 3×16 4×12 6×8

Tell your child: "The greatest common factor for the numbers 36 and 48 is 12."

Ask your child to identify the greatest common factor for the numbers 12 and 24:

1. **Ask your child to write the factors for the number 12:**

    12
    1×12 2×6 3×4

2. **Ask your child to write the factors for the number 24:**

    24
    1×24 2×12 3×8 4×6

3. **Ask your child to find the greatest common factor. They should say it is 12. If they do not, review the factors for each number, selecting the largest factor that both numbers have in common.**

FIND ONLINE

This book's companion website at www.dummies.com/go/teachingyourkidsnew math6-8fd contains a worksheet (the first few rows of which are shown in Figure 8-3) that your child can use to identify the greatest common factor for two numbers. Download and print the worksheet. Help your child solve the first few and then ask them to complete the rest.

6, 10   12,24   8,12

FIGURE 8-3:
A worksheet for finding the greatest common factor.

3,6   5,10   4,16

Understanding the Least Common Multiples

For a given number, a multiple is the result of an integer value times the number. For example, the multiples of 2 are:

2, 4, 6, 8, 10, 12, 14, and so on

Likewise, the multiples of 3 are:

3, 6, 9, 12, 15, 18, and so on

In this section, your child can practice identifying the least common multiple between two numbers.

Say to your child: "A multiple of a number is the result of an integer times the number. Consider, for example, the number 3." Then list the multiples of 3:

3, 6, 9, 12, 15, 18, and so on

Say to them: "Likewise, the multiples of the number 4 are as follows." Then list the multiples of 4:

4, 8, 12, 16, 20, 24, and so on

Say to your child: "There will be times when you must identify the least common multiple for two values. Assume, for example, during the holiday season, you decorate your house with three different types of lights. The first string of lights illuminates every second. The second string illuminates every two seconds and the third every three seconds. By knowing the least-common multiple, you can determine that all three strings of lights will illuminate together every six seconds."

Explain to your child: "If you consider the values 3 and 4, the least (smallest) multiple the two have in common is 12."

3, 6, 9, **12**, 15, 18, 21,

4, 8, **12**, 16, 20, 24,

Ask your child to identify the least common multiple for the numbers 3 and 5:

1. **List the multiples of 3:**

   3, 6, 9, 12, 15, 18, 21

2. **List the multiples of 5:**

   5, 10, 15, 20, 25, 30,

3. **Identify the smallest multiple the two numbers have in common, which in this case is 15.**

FIND ONLINE

This book's companion website at www.dummies.com/go/teachingyourkidsnew math6–8fd contains a worksheet (the first few rows of which are shown in Figure 8-4) that your child can use to identify the least common multiples for two values. Download and print the worksheet. Help your child solve the first few and then ask them to solve the rest.

DEFINITELY IMPRESS YOUR FRIENDS AND FAMILY

As if knowing what is a prime number isn't enough to impress your friends and family, you can definitely "wow" them with your knowledge of Goldbach's Conjecture (*conjecture* is a fancy word for hypothesis) that states, "Every even number is the sum of two prime numbers." For example, consider the following:

$4 = 3 + 1$

$6 = 3 + 3$

$8 = 5 + 3$

$50 = 3 + 47$

$100 = 3 + 97$

3, 5   4,6   7,8

FIGURE 8-4:
A worksheet for
identifying least
common
multiples. 5,6   7,3   4,5

2

Sailing through Sixth-Grade Math

Chapter **9**

Getting Around Shapes

E arly in your child's education, they learned the common shapes, such as squares, triangles, and circles. They may have also learned to calculate the distance around a shape, which is called the *perimeter*, and the space within a shape, which is called the *area*. In this chapter you first review those operations with your child and then perform them with more complex shapes. (Chapter 27 then builds on this knowledge even further by working with 3D shapes to calculate volumes.)

If you have replaced carpet or estimated the amount of paint you need to paint a room, you've calculated the area of a floor or walls. Likewise, if you have placed a fence around your grass or had to calculate the amount of baseboard you need within a room, you have calculated a perimeter. In this chapter, your child learns to do just that.

So, get ready to get in shape, or at least ready to work with them!

Measuring the Perimeter around Shapes

Geometry is the branch of mathematics that deals with space, lines, angles, and shapes. When your child gets to high school, they will take an entire course devoted to geometry, but elementary and middle school get the ball rolling with some basics. In this section, your child gets started with simple geometry. Specifically, they learn to calculate the *perimeter*, or distance around a shape.

Say to your child: "All shapes have a few things in common. They have a distance around them and an area within them. Now, you will learn to calculate a shape's perimeter — the distance around the shape."

Ask your child to consider the following rectangle:

1. **Tell your child: "To calculate the perimeter of a rectangle, you sum up the lengths of each of its sides."**

2. **Have your child add up the side lengths:**
   $5 + 3 + 5 + 3 = 16$

3. **Say to your child: "In this case, the perimeter, or distance around the rectangle, is 16."**

TIP

To help your child understand the concept of a perimeter, pick a square or rectangular room in your house. Explain that you may want to replace the baseboard. Use a tape measure to determine the amount of baseboard you would need.

Ask your child to consider the following square:

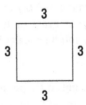

1. **Remind your child: "A square is a special type of rectangle, one for which each side has the same length."**

2. **Have your child add up the side lengths:**
   $3 + 3 + 3 + 3 = 12$

3. **Say to your child: "In this case, the square's perimeter is 12."**

Ask your child to consider the following triangle:

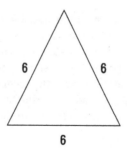

1. Explain to your child: "To calculate the perimeter of a triangle, you add up the lengths of each side."

2. Have your child add the lengths of the triangle's sides:
$$6 + 6 + 6 = 18$$

3. Say to your child: "For this triangle, the perimeter is 18."

Ask your child to compute the perimeters of the following shapes:

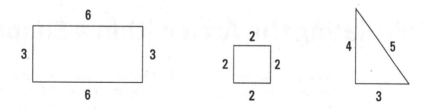

Your child should get:

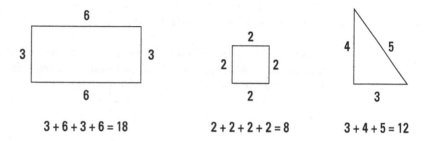

$$3 + 6 + 3 + 6 = 18 \qquad 2 + 2 + 2 + 2 = 8 \qquad 3 + 4 + 5 = 12$$

If they do not, review the shapes with your child and help them add the lengths of the sides.

FIND
ONLINE

This book's companion website at www.dummies.com/go/teachingyourkidsnew math6-8fd contains a worksheet (part of which is shown in Figure 9-1) that your child can use to calculate the perimeters of common shapes. The worksheet consists of two pages. The first lets your child build confidence calculating perimeters that use small numbers. The second page increases the difficulty by increasing the numbers. Download and print the worksheet. Help your child with the first few and then ask them to complete the rest.

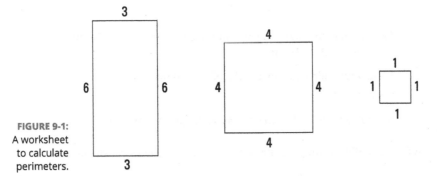

FIGURE 9-1:
A worksheet
to calculate
perimeters.

Calculating the Area within a Shape

In the previous section, your child learned to calculate the perimeter, or distance around rectangles, squares, and triangles. In this section, they learn to calculate the area inside of the shape. You will again now work with common shapes. In a later chapter you will perform similar operations with more complex shapes.

Say to your child: "You have learned to calculate the distance around a shape, which is called the *perimeter*. In this section, you learn how to calculate the space within a shape, which is called the *area*."

Tell your child: "If you want to know how much baseboard you need around a room, you calculate the perimeter of the room. If, instead, you want to change the carpet or flooring inside a room, you calculate the area."

Ask your child to consider the following rectangle:

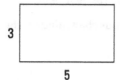

1. Explain to your child: "To calculate the area of a rectangle, you multiply the rectangle's length times its width."

2. Have your child multiply the length and width:

$$\begin{array}{r} 5 \\ \times\ 3 \\ \hline 15 \end{array}$$

3. Say to your child: "For this rectangle, the area inside the shape is 15."

Ask your child to calculate the area for the following rectangle:

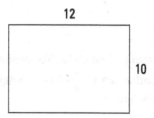

They should get 120. If they don't, help them review the multiplication of the rectangle's length and width.

TIP

To help your child understand the concept of area, pick a square or rectangular room in your house. Tell your child that you want to know how much carpet the room requires. In other words, you want to know the room's area. Use a tape measure to determine the room's length and width, and then calculate the area.

Ask your child to calculate the area of the following square:

1. Remind your child: "A square is a special type of rectangle for which the length and width are the same. To calculate the area of a square, you multiply the two sides."

2. Have your child multiply the square's sides:

$$\begin{array}{r} 4 \\ \times\,4 \\ \hline 16 \end{array}$$

3. Say to your child: "For this square, the area is 16."

Ask your child to consider the following triangle:

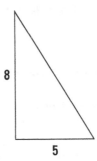

1. Explain to your child: "To calculate the area of a triangle, you use the equation, area = $\frac{1}{2}$ × base × height. In this case, the triangle's base is 5 and the height is 8."

2. Have your child solve the equation:

$$\frac{1}{2} \times \text{base} \times \text{height} =$$

$$\frac{1}{2} \times 5 \times 8 = 20$$

3. Say to your child: "For this triangle, the area is 20."

Ask your child to calculate the areas of the following shapes:

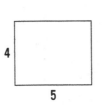

They should get:

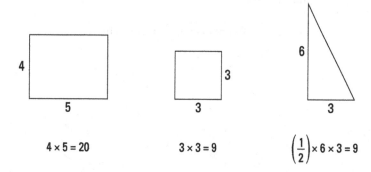

$$4 \times 5 = 20 \qquad 3 \times 3 = 9 \qquad \left(\frac{1}{2}\right) \times 6 \times 3 = 9$$

If they do not, review the shapes with them and help them perform the multiplications.

FIND ONLINE

This book's companion website at www.dummies.com/go/teachingyourkidsnew math6–8fd contains a worksheet (part of which is shown in Figure 9-2) that your child can use to calculate the area for common shapes. Download and print the worksheet. Help your child solve the first few and then ask them to complete the rest. This worksheet also contains two pages. The first page contains shapes with smaller values and the second uses larger numbers.

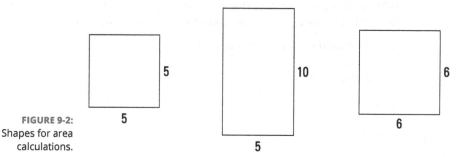

FIGURE 9-2:
Shapes for area calculations.

Working with Circles

In the previous sections, your child learned to calculate the perimeter and area for rectangles, squares, and triangles. In this section, they calculate the same for circles.

TIP

When your child works with circles, they must perform math operations using the value 3.14 for π (pi). You can let your child use a calculator to perform the multiplication.

Present the following circle to your child:

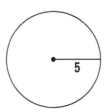

5

Explain to your child: "A circle, like other shapes, has a distance around (which we call the circle's *circumference*) and space inside (an area). To calculate the distance around and the area inside of a circle, you must know the circle's *radius,* which is the distance from the center of the circle to the outer edge. The 5, in this case, is the circle's radius, which helps you determine the circle's size."

Say to your child: "To calculate the circumference of a circle, you use a special value called *pi* (pronounced 'pie') that you represent using the symbol π and for which you use the value 3.14."

1. Tell your child: "To calculate the circumference, or distance around the circle, you use the equation, circumference $= 2 \times \pi \times$ radius .

2. Have your child solve the equation:
   $2 \times \pi \times$ radius $=$
   $\quad 2 \times 3.14 \times 5 = 31.4$

3. Say to your child: "For this circle, the circumference is 31.4."

Help your child calculate the circumference for the following circle:

3

They should get $2 \times \pi \times 3 = 18.84$. If they don't, help them work through the equation.

Just as a circle has a circumference around it, a circle also has a space inside of it — the circle's area. You will now show your child how to calculate the area, the formula of which again uses pi and the circle's radius.

1. Explain to your child: "To calculate the area of a circle, you use the equation, area = $\pi \times$ **radius** $\times$ **radius**."

2. Have your child solve the equation:
   $\pi \times \text{radius} \times \text{radius} =$
   $\qquad 3.14 \times 4 \times 4 = 30.24$

3. Say to your child: "For this circle, the area is 30.24."

Help your child compute the area for the following circles:

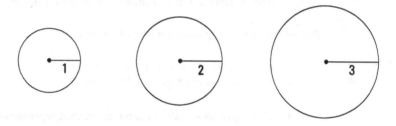

They should get:

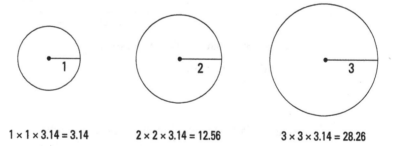

$1 \times 1 \times 3.14 = 3.14$ $\qquad$ $2 \times 2 \times 3.14 = 12.56$ $\qquad$ $3 \times 3 \times 3.14 = 28.26$

FIND ONLINE

This book's companion website at www.dummies.com/go/teachingyourkidsnew math6–8fd contains a worksheet (part of which is shown in Figure 9-3) that your child can use to find the perimeter and area of circles. Download and print the worksheet. Help your child solve the first few and then ask them to complete the rest.

FIGURE 9-3:
Circles for
perimeter and
area calculations.

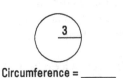

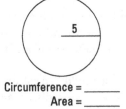

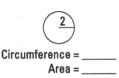

Circumference = _____
Area = _____

Circumference = _____
Area = _____

Circumference = _____
Area = _____

Working through Word Problems with Perimeters and Areas

In this section, your child completes word problems. This time, the problems require them to calculate the perimeter or area of a shape.

Present the following word problem to your child:

> Bill and his dad want to sod the backyard with grass. The yard is 50 feet long and 20 feet wide. How much grass will they need?

1. **To start, have your child draw a picture that represents the problem:**

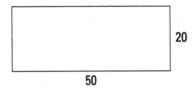

2. **Have your child calculate the yard's area:**

$$Area = length \times width$$
$$= 50 \times 20$$
$$= 100$$

Present the following word problem to your child:

> Mary wants to place a wallpaper border around her bedroom. The room is rectangular with walls 8 feet wide by 12 feet long. How much wallpaper does Mary need?

1. **To start, have your child draw a picture that represents the problem:**

```
          12
   ┌──────────────┐
   │              │
 8 │              │ 8
   │              │
   └──────────────┘
          12
```

2. **Have your child calculate the room's perimeter:**

$$12 + 8 + 12 + 8 = 40$$

Present the following word problem to your child:

> Juan's family wants to place a circular fire pit in their backyard. The radius of the pit (the circle) will be 5 feet. What are the pit's circumference and area?

1. **To start, have your child draw a picture that represents the problem:**

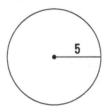

2. **Have your child find the circumference:**

$$\begin{aligned} \text{Perimeter} &= 2 \times \pi \times \text{radius} \\ &= 2 \times 3.14 \times 5 \\ &= 31.4 \end{aligned}$$

3. **Have your child find the area:**

$$\begin{aligned} \text{Area} &= \pi \times \text{radius} \times \text{radius} \\ &= 3.14 \times 5 \times 5 \\ &= 78.75 \end{aligned}$$

Chapter **10**

Working with Numbers Less than Zero

positive attitude is important — being positive is a good thing. People say that you should avoid being negative, but what does it really mean to be negative? In this chapter, your child finds out!

You all learn to count at a young age. Counting, for the most part, is pretty easy — the numbers always get bigger. Well, that's true until you're introduced to negative numbers. Your child may have seen negative numbers for very cold temperatures or when watching you trying to balance your budget. Now, your child gets to work with them on number lines and to solve math problems that contain them.

So, get ready to go negative and to think about the question, "How low can you go?"

Understanding Negative Numbers

Until now, your child has started counting with either 0 or 1, and numbers have only gotten larger when they count. In this section, your child learns to work with negative numbers.

Say to your child: "For years, you have started counting with either 0 or 1. You have learned that numbers can increase infinitely, meaning that no matter how large a number is, you can always add one to it."

Explain to your child: "Until now, numbers on your number line have only gotten larger. You call numbers that are greater than 0 *positive* numbers."

Say to your child: "It turns out that numbers can be less than 0, or *negative*. You represent negative numbers (which are located on the number line to the left of 0) by preceding each number with a negative sign, as shown here."

Explain to your child: "A real-world example of negative numbers is temperature. During January, in Minnesota, the weather can get very cold, with temperatures less than 0, such as −5 or −10 degrees, or even colder."

Say to your child: "You have learned to use a number line to subtract one number from another."

$5 - 4 =$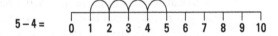

Tell your child: "Using a number line with positive and negative numbers, consider the following."

$5 - 6 =$

Explain to them: "In this case, the answer becomes −1, as shown here."

$5 - 6 =$

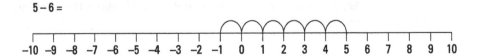

Say to your child: "In a similar way, consider the following."

$$3 - 7 =$$

Tell them: "Using the number line, your result becomes −4."

$$3 - 7 =$$

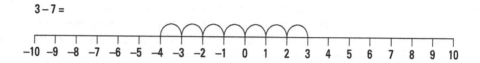

Ask your child to use a number line to solve the following expressions:

$$2 - 5 = \qquad 4 - 6 = \qquad 7 - 10 = \qquad 5 - 9 =$$

They should get:

$$2 - 5 = -3 \qquad 4 - 6 = -2 \qquad 7 - 10 = -3 \qquad 5 - 9 = -4$$

If they do not, help them use the number line to perform the subtractions.

FIND ONLINE

This book's companion website at www.dummies.com/go/teachingyourkidsnewmath6-8fd contains a worksheet (the first few rows of which are shown in Figure 10-1) that your child can use to subtract numbers that result in negative numbers. Download and print the worksheet. Help your child solve the first few and then ask them to complete the rest.

$1 - 2 =$	$2 - 3 =$	$3 - 4 =$	$0 - 1 =$
$2 - 4 =$	$5 - 6 =$	$4 - 7 =$	$1 - 4 =$
$3 - 6 =$	$2 - 4 =$	$0 - 5 =$	$4 - 8 =$

FIGURE 10-1: A worksheet for subtracting numbers with negative results.

PASTA AND NEGATIVE NUMBERS — THANK THE CHINESE

Archeologists believe the Chinese created the first pasta noodles about 3,000 years ago. Not long after that, according to the book *The Nine Chapters on the Mathetical Art*, the Chinese introduced the concept of negative numbers. Unfortunately, thousands of years later, negative numbers still weren't positively received by mathematicians. Thankfully, around 1700, an English mathematician named John Wallis created a number line giving new meaning to negative numbers!

Comparing Positive and Negative Numbers

In the previous section, your child learned about negative numbers. In this section, they compare positive and negative numbers.

Say to your child: "You know how to use the less than (<), equal (=), and greater than (>) symbols to compare numbers. In this section, you use the symbols to compare positive and negative numbers."

Ask your child to use <, =, and > to compare the following numbers. If necessary, allow your child to use the number line as they get started.

–3	0	–4	–1	3	1
4	–4	–4	5	7	–7
–1	0	–1	–1	–3	–2

Your child should get the following:

–3	<	0	–4	<	–1	3	>	1
4	>	–4	–4	<	5	7	>	–7
–1	<	0	–1	=	–1	–3	<	–2

If your child misses a problem, use the following number line to help them correct the error.

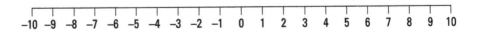

This book's companion website at www.dummies.com/go/teachingyourkidsnew math6–8fd contains a worksheet (the first few rows of which are shown in Figure 10-2) that your child can use to compare numbers. Download and print the worksheet. Help your child solve the first few and then ask them to complete the rest.

10	−3		−3	5		−7	7		−7	−7
−1	0		0	−4		4	4		1	−1
−3	4		4	−7		5	0		1	1

FIGURE 10-2: A worksheet for comparing positive and negative numbers.

Adding Negative Numbers

Once your child has a good handle on positive and negative numbers, they can move on to adding negative numbers.

Present the following expression to your child:

$$4+(-3)=$$

Explain to your child: "When you add a negative number, you change the sign to subtraction, as shown here."

$$4+(-3)=$$
$$4-3=1$$

Ask your child to solve the following expressions:

$$0+(-2)= \qquad 3+(-4)= \qquad -3+(-2)=$$

1. **Have your child change the sign of the operators as shown here:**

$$0-2= \qquad 3-4= \qquad -3-2=$$

2. **Have your child solve the expressions, using a number line if necessary:**

$$0-2=-2 \qquad 3-4=-1 \qquad -3-2=-5$$

This book's companion website at www.dummies.com/go/teachingyourkidsnew math6–8fd contains a worksheet (the first few rows of which are shown in Figure 10-3) that your child can use to add negative numbers. Download and print the worksheet. Help your child solve the first few and ask them to complete the rest.

$$10 + (-3) = \qquad -3 + 5 = \qquad 7 + (-7) = \qquad 7 + (-2) =$$

$$0 + (-1) = \qquad 0 + (-4) = \qquad 4 + (-4) = \qquad 1 + (-1) =$$

FIGURE 10-3:
A worksheet for
adding negative
numbers.

$$3 + (-4) = \qquad 4 + (-7) = \qquad 5 + 0 = \qquad 2 + (-1) =$$

Subtracting Negative Numbers

Just as you can add negative numbers (covered in the previous section), you can also subtract negative numbers. This section walks you through how to do just that.

Say to your child: "As you learned, the process of adding a negative number is actually the same as subtracting the corresponding positive number."

Tell your child: "Next, you will learn to subtract a negative number. As you will find, subtracting a negative number is actually the same as adding the corresponding positive number."

Ask your child to consider the following expressions:

$$2 - (-5) = \qquad 3 - (-3) = \qquad -7 - (-5) =$$

1. **Have your child change the signs of each expression as follows:**

$$2 - (-5) = \qquad 3 - (-3) = \qquad -7 - (-5) =$$
$$2 + 5 = \qquad 3 + 3 = \qquad -7 + 5 =$$

2. **Have your child solve the expressions, using a number line as necessary:**

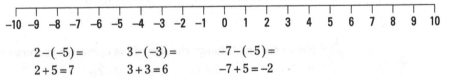

$$2 - (-5) = \qquad 3 - (-3) = \qquad -7 - (-5) =$$
$$2 + 5 = 7 \qquad 3 + 3 = 6 \qquad -7 + 5 = -2$$

**FIND
ONLINE**

This book's companion website at www.dummies.com/go/teachingyourkidsnew math6-8fd contains a worksheet (the first few rows of which are shown in Figure 10-4) that your child can use to subtract negative numbers. Download and print the worksheet. Help your child solve the first few and then ask them to complete the rest.

$10 - (-3) =$	$-3 - 5 =$	$7 - (-7) =$	$7 - (-2) =$

FIGURE 10-4:
A worksheet for
subtracting
negative
numbers.

$0 - (-1) =$	$0 - (-4) =$	$4 - (-4) =$	$1 - (-1) =$

$3 - (-4) =$	$4 - (-7) =$	$5 - 0 =$	$2 - (-1) =$

Multiplying Negative Numbers

In the preceding sections, your child learned to add and subtract negative numbers. Once they have that down, they can learn to multiply them.

Say to your child: "In this section, you learn to multiply negative numbers. You learn that the multiplication process does not change; the new issue is determining the sign (positive or negative) of the result."

Ask your child to consider the following expression:

$$3 \times (-4) =$$

Say to your child: "To start, you simply perform the multiplication $3 \times 4 = 12$. Then, you determine the sign of the result by counting how many numbers are negative. If the count is even, the result is positive. If the count is odd, the result is negative."

Explain to your child: "In this case, only one number is negative (an odd), which means the result is negative."

$$3 \times (-4) = -12$$

Ask your child to solve the following expressions:

$$(-5) \times 2 = \qquad 3 \times (-5) = \qquad (-2) \times (-2) =$$

1. **Have your child count the number of values that are negative to determine the result's sign:**

$$(-5) \times 2 = - \qquad 3 \times (-5) = - \qquad (-2) \times (-2) =$$

2. **Have your child complete the multiplication:**

$$(-5) \times 2 = -10 \qquad 3 \times (-5) = -15 \qquad (-2) \times (-2) = 4$$

When multiplying two negatives, the result is always positive.

REMEMBER

This book's companion website at www.dummies.com/go/teachingyourkidsnew math6-8fd contains a worksheet (the first few rows of which are shown in Figure 10-5) that your child can use to multiply negative numbers. Download and print the worksheet. Help your child solve the first few and then ask them to complete the rest.

$10 \times (-3) =$ $-3 \times 5 =$ $7 \times (-7) =$ $7 \times (-2) =$

$0 \times (-1) =$ $0 \times (-4) =$ $4 \times (-4) =$ $1 \times (-1) =$

$3 \times (-4) =$ $4 \times (-7) =$ $5 \times 0 =$ $2 \times (-1) =$

FIGURE 10-5:
A worksheet for multiplying negative numbers.

Dividing Negative Numbers

As your child will find, for division, the process of determining the sign of the result is the same as for multiplication.

Say to your child: "In the previous section, you learned to multiply negative numbers. Now you will learn to divide them. As you will see, the process is very similar."

Ask your child to consider the following expression:

$10 \div (-5) =$

Explain to your child: "When you divide a negative number, the division process does not change. To determine the sign of the result, you again count how many of the numbers are odd. If the count is odd, the result is negative. If the count is even, the result is positive."

Say to your child: "In this case, because there is only one negative number, the result is negative."

$10 \div (-5) = -2$

Ask your child to solve the following expressions:

$-6 \div 2 =$ $-8 \div (-4) =$ $10 \div (-5) =$

1. **Have your child determine the sign of the result:**

$-6 \div 2 = \quad - \qquad -8 \div (-4) = \qquad 10 \div (-5) = \quad -$

2. **Have your child complete the division:**

$-6 \div 2 = -3 \qquad -8 \div (-4) = 2 \qquad 10 \div (-5) = -2$

FIND ONLINE

This book's companion website at www.dummies.com/go/teachingyourkidsnew math6–8fd contains a worksheet (the first few rows of which are shown in Figure 10-6) that your child can use to divide negative numbers. Download and print the worksheet. Help your child solve the first few and then ask them to complete the rest.

$9 \div (-3) =$	$-15 \div 5 =$	$7 \div (-7) =$	$8 \div (-2) =$
$10 \div (-1) =$	$12 \div (-4) =$	$4 \div (-4) =$	$1 \div (-1) =$
$16 \div (-4) =$	$28 \div (-7) =$	$5 \div (-1) =$	$2 \div (-1) =$

FIGURE 10-6: A worksheet for dividing negative numbers.

Understanding the Additive Inverse

As your child progresses in math, there are times when they must learn new terms. This is one of those times. In this section, your child learns the term *additive inverse*.

Say to your child: "If you examine a number line, you know 1 is the same distance from 0 as −1, as are 2 and −2."

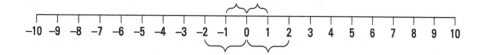

Explain to your child: "The term *additive inverse* describes the positive or negative value you must add to a number to get 0 as a result."

Say to your child: "For example, the additive inverse of 5 is −5."

$$5 + (-5) = 0$$

Tell them: "Likewise, the additive inverse of −3 is 3."

$$-3 + 3 = 0$$

Ask your child to find the additive inverse for the following:

7	−6	9

They should get:

−7	6	−9

If your child does not, use a number line to ask them what number they must add to a given value to get 0.

Understanding the Absolute Value

Time for another new term: *absolute value*.

Say to your child: "When you work with negative numbers, there may be times when you want to force the result to always be positive. For example, assume you forget to stop for gas and your car stalls somewhere on a dark desert highway. Fortunately, a stranger emerges and tells you there is a gas station three miles back. When dealing with distances, for example, you use positive numbers. You don't say, there's a gas station in −3 miles. For such cases when you only want positive values, you use the *absolute value*."

Say to them: "You represent the absolute value using two vertical bars. Consider the following."

$\lvert -3 \rvert = 3$	$\lvert 2 \rvert = 2$	$\lvert 1 - 3 \rvert = 2$	$\lvert -5 \rvert = 5$

Tell your child: "The result of an absolute value is always positive."

Ask your child to calculate the absolute value of the following:

$\lvert -7 \rvert =$	$\lvert 3 - 4 \rvert =$	$\lvert -7 - 3 \rvert =$	$\lvert -1 \times 5 \rvert =$

Your child should get:

$$|-7| = 7 \qquad |3-4| = 1 \qquad |-7-3| = 10 \qquad |-1 \times 5| = 5$$

The result of an absolute value is always positive.

Charting Negative Values

When your child works with negative numbers, there may be times when they must chart negative values. In this section, your child learns to do just that.

Say to your child: "In previous grades, you learned to chart data. The following chart, for example, shows the January temperatures for a week in Phoenix, Arizona."

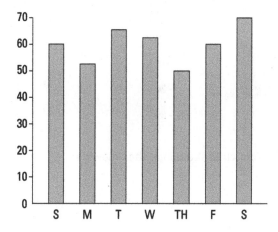

Say to your child: "As you can see, even during January, the days in Phoenix are warm, and so the chart starts with 0."

Tell your child: "The following data, in contrast, shows the low temperatures for a week in January for St. Paul, Minnesota."

Day	Low Temperature
Sunday	5
Monday	10
Tuesday	–3

Day	Low Temperature
Wednesday	–10
Thursday	–10
Friday	5
Saturday	10

To chart this data, you need a chart with both negative and positive values:

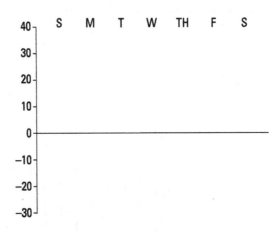

Your child should get the following:

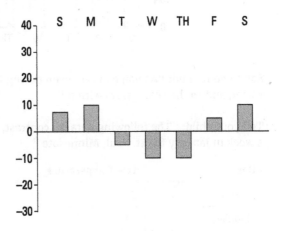

Ask your child to use the following chart to graph a company's profit-and-loss data for a week:

Day	Result
M	50
T	100
W	−50
TH	−100
F	50

Your child should get the following:

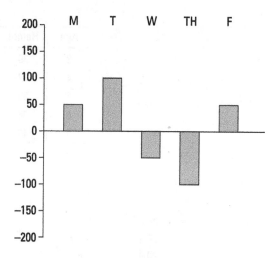

Your child has learned how to graph positive numbers, such as that shown:

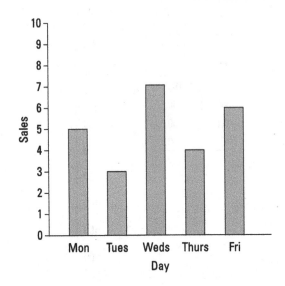

Using the following graph, ask your child to graph the ages and heights shown:

Age	Height
3	30
4	32
5	35
6	38
7	40

Your child should get:

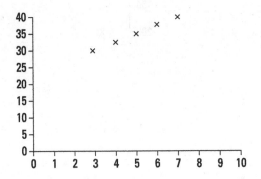

As you just learned with the temperature chart, there will be times when your child must graph negative numbers.

When you chart data, you will often refer to the horizontal chart axis as the x-axis and the vertical axis as the y-axis:

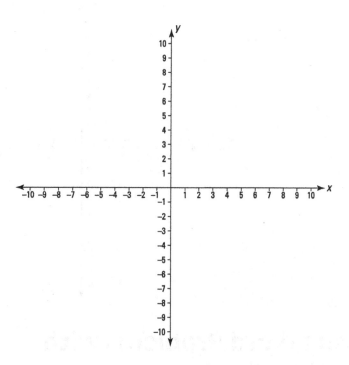

Note that each axis, like a number line, has positive and negative numbers. Using the chart, help your child plot the following data points:

X values	Y values
–5	–3
–3	–1
–1	1
1	3
3	5

Have your child draw a line through the data points. They should get the chart shown:

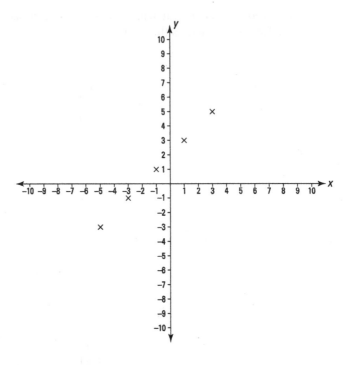

Solving Word Problems with Negative Numbers

In this section, your child gets to practice solving word problems with negative numbers.

Present the following problem to your child:

> Mary lives in New York where the January temperature is 32 degrees. Sally lives in Maine where the January temperature is 15 degrees. How much warmer is New York than Maine?

Your child should identify the following expression:

> Temperature Difference = 32 – 15

Your child should state that New York is 17 degrees warmer than Maine.

Next, present the following problem to your child:

> Billy lives in Phoenix, Arizona, where the January temperature is 70 degrees. Mary lives in Anchorage, Alaska, where the January temperature is –10 degrees. How much warmer is Phoenix than Anchorage?

Explain to your child, "To determine the temperature difference, you must subtract the temperature in Anchorage from the temperature in Phoenix." Help your child create the following expression:

> Temperature difference = 70 – (–10)

Have your child perform the subtraction. They should find that Phoenix is 80 degrees warmer than Anchorage.

Present the following problem to your child.

> Mary and Alise are playing a video game. Mary has 40 points and Alise has 15. By how many points is Mary leading?

Your child should identify the following expression:

> Lead = 40 – 15 = 25

Next, present the following problem to your child:

Tanya and Juan are playing the same video game. Tanya's current score is 30. Juan's current score is –20. How many points does Juan need to earn to catch up with Tanya?

Help your child create the following expression:

> Difference = 30 – (–20)

Have your child solve the expression. They should get 50.

IN THIS CHAPTER

» **What are statistics?**

» **Calculating the average value (mean)**

» **Identifying the maximum and minimum values**

» **Sorting data**

» **Making sense of the median and mode**

» **What charts tell you about data**

Chapter **11**

Starting with Simple Statistics

S tatistics. Just the mention of the word can strike fear (or boredom) into the average person or cause readers of a certain book to skip to the next chapter! Yeah, it's okay if you look again at this chapter's title — that is, as long as you promise to keep reading.

Statistics is about data. Specifically, statistics is the process of collecting and analyzing data. If that sounds complicated and hard, relax! You're going to look at simple statistics such as minimum and maximum values, the average value (which statisticians call the *mean*), the range between minimum and maximum values, and so on.

Kids start working with statistics early in their math careers. By second grade, they start creating graphs and bar charts for given data. So, take a deep breath and let's get started. I think you will find statistics fun to teach to your child. In addition, you will likely impress your friends the next time they tell you a fact from the news and you say, "I think outliers are causing a large variance in the result!" So, get out your plastic pocket protector and get ready to do some "stats."

It All Starts with Data

Statistics is the math associated with collecting and analyzing data. The first part of statistics, therefore, is about collecting data. To collect data, you ask questions similar to the following and record (write down) the results:

>> How much did you pay for gas?

>> How many pages did you read?

>> How much did you spend on groceries?

>> How many hours did you study?

>> For whom do you plan to vote?

If you ask several people one of these questions, you have data you can begin to analyze, and that makes you a statistician. Cool, right?

Calculating the Average (Mean) Value

Say to your child: "Today you're going to analyze data and perform statistical operations. You're going to start with the average value, which provides you with a way to quickly summarize your data. When you perform statistics, you call the average value the mean."

Present the following quiz scores to your child:

Bill	8
Mary	10
Javiar	9

1. Tell your child: "You have 3 quiz scores. To compute the average score, you first add up the quiz scores."

$$\begin{array}{r} 8 \\ +10 \\ \underline{+\ 9} \\ 27 \end{array}$$

2. Tell them: "Next, you divide your sum by the number of quiz scores to get the average."

$$3\overline{)27}$$
$$\underline{-27}$$
$$0$$

with quotient 9 above.

3. Say to your child: "In this case, the average quiz score was 9."

Present the following data to your child:

> Marta weighs 80 pounds.
>
> Joey weighs 90 pounds.

Ask your child: "What is the average weight?"

1. To start, have your child add up the weights:

$$\begin{array}{r} 80 \\ +90 \\ \hline 170 \end{array}$$

2. Have your child divide the sum by the number of data values, which, in this case, is 2:

$$\begin{array}{r} 85 \\ 2\overline{)170} \\ -16 \\ \hline 10 \\ -10 \\ \hline 0 \end{array}$$

3. Tell your child: "In this case, the average weight is 85 pounds."

Present the following data to your child:

> Students in the class read the following number of books:
>
> Thomas read 2 books.
>
> Malia read 3 books.
>
> Juan read 5 books.
>
> Shanae read 6 books.

Ask your child to calculate the average number of books the students read.

1. Have your child sum (add up) the data values:

$$\begin{array}{r} 2 \\ +3 \\ +5 \\ +6 \\ \hline 16 \end{array}$$

2. **Have your child divide the sum by the number of data values, which, in this case, is 4:**

$$4\overline{)\,16} \;\;\begin{array}{c}4\\-16\\\hline 0\end{array}$$

3. **Ask your child: "What was the average number of books read?"**

TIP

Explain to your child that the "average value" is also called the "mean value." They can use the words *average* and *mean* interchangeably.

Present the following data sets to your child and ask them to calculate the average value for each:

Family Size	Shoe Size	Age
3	5	5
4	6	15
2	8	10
	9	20
		15

Your child should get the following:

$$\begin{array}{l}3\\4\\2\\\hline 9\end{array}\qquad 3\overline{)\,9}\;{}^{3}$$

$$\begin{array}{l}5\\6\\8\\9\\\hline 28\end{array}\qquad 4\overline{)\,28}\;{}^{7}$$

$$\begin{array}{l}5\\15\\10\\20\\15\\\hline 65\end{array}\qquad 5\overline{)\,65}\;{}^{13}$$

FIND
ONLINE

This book's companion website at www.dummies.com/go/teachingyourkidsnew math6-8fd provides a worksheet (the first few rows of which are shown in Figure 11-1) that your child can use to solve average values. Download and print the worksheet. Help your child solve the first few and then ask them to complete the rest.

DID YOU KNOW?

Following are some fun stats to impress your friends:

- The oldest (max) living person was 122 ½ years old.
- The average IQ is 100 — Einstein's IQ is estimated to be 200 to 205.
- The team with the most World Series wins (mode value) is the New York Yankees.
- The youngest (min) President was John F. Kennedy at age 43.
- The median age in the United States is 38.

1	10	11
2	20	12
3	30	13
4	40	14
5	50	15

FIGURE 11-1: An average value worksheet.

Finding the Smallest and Largest Values (Minimum and Maximum)

When you analyze data, you often want to know the minimum (smallest) and maximum (biggest) values in the data. In this section, you teach your child to do just that.

Say to your child: "When you analyze data, you often want to know the smallest and biggest values in the data. You call the smallest value the minimum and the biggest value the maximum."

Present the following data to your child:

Test Scores
80
70
30
90
45

Ask your child to find the minimum value. They should say 30. If they do not, help them review the data.

Ask your child to find the maximum value. This time, they should say 90. If they do not, review the data with them and point out the largest value.

Present the following data sets to your child and ask them to find the minimum and maximum value for each:

Family Size	Shoe Size	Age
3	5	5
4	6	15
2	8	10
	9	20
		15

Your child should get:

Family Size	Shoe Size	Age
3	5	5
4	6	15
2	8	10
	9	20
		15
Minimum 2	Minimum 5	Minimum 5
Maximum 4	Maximum 9	Maximum 20

FIND ONLINE

This book's companion website at www.dummies.com/go/teachingyourkidsnew math6-8fd provides a worksheet (the first few rows of which are shown in Figure 11-2) that your child can use to identify minimum and maximum values within data. Download and print the worksheet. Help your child solve the first few and then ask them to complete the rest.

1	100	111
0	20	102
3	300	103
4	40	114
7	504	105

FIGURE 11-2:
A minimum and maximum value worksheet.

Min:

Max:

Min:

Max:

Min:

Max:

Bringing Some Order to Data (Sorting)

When your child works with data, it is often convenient to sort the data from smallest to biggest. In this section, you teach your child to do just that.

Say to your child: "Sorting data is the process of putting the data in order, normally from smallest to biggest."

Present the following age data to your child:

Ages
5
10
1
3
7

Explain to your child: "You are going to sort the data by rewriting the numbers in order, from smallest to biggest."

1. **Ask your child to find the smallest value in the list and to write it in a new list, crossing out the number in the first list:**

Ages	Sorted List
5	1
10	
4	
3	
7	

2. **Have your child find the next smallest number in the list, repeating this process:**

Ages	Sorted List
5	1
10	3
~~1~~	
~~3~~	
7	

3. Have your child repeat this process for the remaining values:

Ages	Sorted List
~~5~~	1
~~10~~	3
~~1~~	5
~~3~~	7
~~7~~	10

Explain to your child: "Using the sorted data, you can easily identify the minimum and maximum values in the list."

Ask your child to sort the following values:

10

30

15

20

They should get:

~~10~~	10
~~30~~	15
~~15~~	20
~~20~~	30

If your child does not sort the values correctly, review the list with them and move the smallest values to the new list one at a time.

This book's companion website at www.dummies.com/go/teachingyourkidsnew math6–8fd provides a worksheet (the first few rows of which are shown in Figure 11-3) that your child can use to sort data. Download and print the worksheet. Help your child solve the first few and then have them complete the rest.

1	100	111
0	20	102
3	300	103
4	40	114
7	504	105

FIGURE 11-3: A data sorting worksheet.

Finding the Value That Occurs Most Often in Data (The Mode)

When you analyze data, there are times when you want to know the value that appears most often in the list; this value is called the *mode*. In this section, you teach your child how to identify the mode.

Say to your child: "When you analyze data, you are sometimes asked to identify the value that occurs most often in the data. You call that value the mode. It tells you the most common data value."

Present the following data to your child:

Ages

8

7

6

7

7

6

5

Explain to your child, "Many statistical operations require that you first sort the data." Have your child sort the data as just discussed. They should get:

Ages
5
6
6
7
7
7
8

Ask your child to look at the data and determine which value occurs most often. Your child should say 7, as it appears 3 times. If they do not, help them count (and maybe write down) how often each value appears.

Ask your child: "What do you call the data value that appears most often within your data set?" You may need to remind your child that the most commonly occurring value is called the mode. It's not a term that often comes up in everyday conversation, or even in math.

Present the following data to your child:

Books Read
4
3
2
3
3
2
1

Again, have your child begin by sorting the data. They should get:

Books Read

1
2
2
3
3
3
4

Ask your child to identify the mode. This time, they should identify the value 3, which happens to appear three times in the list.

FIND ONLINE

This book's companion website at `www.dummies.com/go/teachingyourkidsnew math6-8fd` provides a worksheet (the first few rows of which are shown in Figure 11-4) that your child can use to calculate the mode. Download and print the worksheet. Help your child solve the first few and then have them complete the rest.

1	100	111
0	200	111
3	300	103
4	400	114
3	400	105

FIGURE 11-4: A worksheet for calculating the mode.

Finding the Middle Value in Data (The Median)

Another statistic you often must identify when you analyze data is the middle value, or *median*. After you identify the median, you find that half the values in the list are bigger than the median and half are smaller. In this section, you teach your child to identify the median value.

Say to your child: "You have learned to identify the minimum and maximum data values in a data set. Now you will learn to identify the value that falls in the middle of the list; that value is called the median."

Present the following quiz–score data to your child:

Quiz Scores

8
3
10
9
5

1. Explain to your child: "The median value is the middle value in your sorted list. To start, you must first sort your list from smallest to biggest."

Quiz Scores	Sorted List
8	3
3	5
10	8
9	9
5	10

2. Ask your child to identify the middle value from the sorted list. Your child should say 8. If they do not, help them review the list to find the middle value.

Ask your child to find the median value for the following list of books read:

Books Read

3
7
1
2
5

1. **Have your child sort the list from smallest to biggest:**

Books Read	Sorted List
~~3~~	1
~~7~~	2
~~4~~	3
~~2~~	5
~~5~~	7

2. **Ask your child to find the middle value, which, in this case, is 3.**

FIND ONLINE

This book's companion website at www.dummies.com/go/teachingyourkidsnew math6–8fd provides a worksheet (the first few rows of which are shown in Figure 11-5) that your child can use to calculate the median. Download and print the worksheet. Help your child solve the first few and then have them complete the rest.

1	100	111
0	200	121
3	300	103
4	500	114
7	400	105

FIGURE 11-5: A worksheet for calculating the median.

Identifying the Range of Values

A common question that statisticians ask is about the "range" of data values, which is the difference between the maximum and minimum values. In this section, your child learns to calculate the range of values.

Present the following list to your child:

Weights
50
180
75
85
90

1. **Say to your child, "To start, you must sort the data."**

   They should get:

Weights
50
75
85
90
180

2. **Say to your child: "To find the range of values in your data, you must first find the minimum and maximum values."**

3. **Ask your child to identify the minimum and maximum. They should say 50 and 180. If they don't, help them review the data.**

4. **Ask your child to calculate the range by subtracting the minimum value from the maximum:**

   $$\begin{array}{r} 180 \\ -50 \\ \hline 130 \end{array}$$

5. **Say to your child: "In this case, the range of values, from biggest to smallest, is 130."**

Ask your child to find the range for the following values:

Heights
45
72
60
56

1. **Ask your child to identify the minimum and maximum values. They should say 45 and 72.**

2. Have your child subtract 45 from 72:

$$\begin{array}{r} 72 \\ -45 \\ \hline 27 \end{array}$$

3. Explain to your child that for this data, the range is 27.

FIND ONLINE

This book's companion website at www.dummies.com/go/teachingyourkidsnew math6-8fd provides a worksheet (the first few rows of which are shown in Figure 11-6) that your child can use to identify the range of data. Download and print the worksheet. Help your child solve the first few and then have them complete the rest.

15	210	115
10	220	310
5	330	415
20	114	60
22	250	175

Min:	Min:	Min:
Max:	Max:	Max:
Range:	Range:	Range:

FIGURE 11-6: A worksheet for calculating the range of data.

Understanding Box-and-Whiskers Charts

When statisticians examine data, they often create a chart, called a *box-and-whiskers chart,* that shows the maximum, minimum, and median values, as well as *quartiles* (the ranges at which 25 percent, 50 percent, and 75 percent of the data fall).

Reading a box-and-whiskers chart

In this section, you teach your child to read a box-and-whiskers chart.

Say to your child: "You have learned to calculate many statistics, such as the minimum, maximum, mean, and median value."

Explain to them: "To summarize statistics, statisticians often use a special chart, called a box-and-whiskers chart."

Present the following chart to your child:

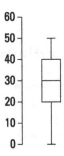

Explain to your child: "This chart is called a box-and-whiskers chart because it has a box and two lines coming from it, which you call whiskers."

Tell your child: "The end of the top line (the top whisker) tells you the maximum value, which, in this case, is 50. Likewise, the end of the bottom whisker line tells you the minimum value, 0."

Say to your child: "The size of the box varies between charts — sometimes it is big and other times it is small, depending on the data. The box tells you the range of values for which 50 percent of the data falls. In this case, you can see that 50 percent (half) of the data values fall in the range 20 to 40."

Say to them: "The middle line in the box tells you your median, or middle value. In this case, the median value is 30."

Present the following box-and-whiskers chart to your child:

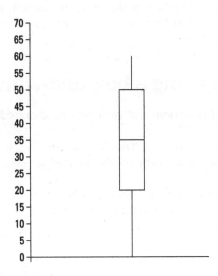

Help your child to identify the following:

>> Maximum value

>> Minimum value

>> Median value

>> The range of values within which 50 percent of your data falls

They should get the following.

>> Maximum value: 60

>> Minimum value: 0

>> Median value: 35

>> The range of values within which 50 percent of your data falls: 20 to 50

Understanding quartiles

If you examine the previous box-and-whiskers charts, you will find that the whiskers coming out of the box are not the same length. Statisticians use the whiskers to identify *quartiles* — the range of values within which the lowest 25 percent of values fall and the range of values within which the upper 25 percent of values fall. Wow! That seems hard! Read on to understand what this means.

You know that the box within the box-and-whiskers chart shows the range of values within which 50 percent of your data falls. That means 50 percent of the data falls outside of the box. Twenty-five percent of the data is bigger, and 25 percent is smaller. The whiskers tell you the range of values for the 25 percent that are bigger, as well as the range of values for the 25 percent that are smaller. Again, depending on the data, the lengths of the two whiskers may differ.

REMEMBER

Box-and-whiskers charts can be a lot for your child to take in; quartiles are a difficult concept. For now, have your child focus on the minimum, maximum, and median values. They should also understand that the box tells them the range of values within which 50 percent of the data falls.

Creating a Histogram

In statistics, you are often interested in how many times each data value occurs, which statisticians call *frequencies*. To visually represent such frequencies, statisticians use a *histogram chart*.

Consider the following histogram chart:

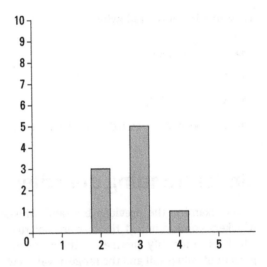

From the histogram, you know the number 2 occurs three times, the number 3 occurs five times, and the number 4 occurs once. You also know that the numbers 1 and 5 did not occur in the data.

Using the following histogram, have your child plot the frequencies of the following data:

Ages
1
1
2
2
2
3
4

Ages
4
5
5
5

They should create the following histogram:

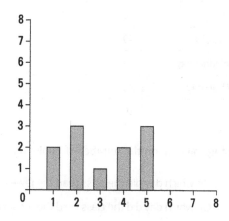

Working Out Statistical Word Problems

By now, you and your child should be getting used to word problems! You may be thinking, "Word problems that require statistics — oh boy!" The good news is that your child should readily solve the following.

Present this word problem to your child:

Tim is 15, Mary is 10, and Billy is 5. What is their average age?

1. **Have your child sum the ages:**

```
  15
+10
+ 5
 30
```

2. Have your child divide the sum by the number of data values, which, in this case, is 3:

$$3\overline{)30} \atop 10$$

Present this word problem to your child:

Each day, the class tracked the number of books the students read:

Day	Number of Books
Monday	30
Tuesday	40
Wednesday	25
Thursday	35
Friday	50

Using the data on the number of books read, find the following:

On which day did the class read the fewest books (the minimum)?

On which day did the class read the most books (the maximum)?

What was the mean number of books read each day?

What was the range of books read (the maximum minus the minimum)?

Present the following data to your child:

10

10

20

30

30

30

40

40

Identify the following:

>> The mean value

>> The minimum value

>> The maximum value

>> The median

>> The mode

>> The range of values

Your child should get

>> The mean value: 27.5

>> The minimum value: 10

>> The maximum value: 50

>> The median value: 30

>> The mode value: 30

>> The range of values: (50 − 10) = 40

Present the following data to your child:

10.3

20.5

30.3

30.7

40.4

40.4

50.1

Identify the following:

>> The average value

>> The minimum value

>> The maximum value

>> The median

>> The mode

>> The range of values

Your child should get

>> The average value: 31.81

>> The minimum value: 10.3

>> The maximum value: 50.1

>> The median: 30.7

>> The mode: 40.4

>> The range of values: (50.1 – 10.3) = 39.8

Present the following box-and-whiskers chart to your child:

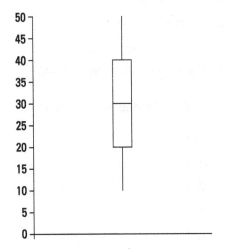

Ask your child to identify the following:

>> Median value

>> Minimum value

>> Maximum value

>> Range within which 50% of the values fall

>> Range of the top quartile

>> Range of the bottom quartile

Your child should get:

>> Median value: 30

>> Minimum value: 10

>> Maximum value: 50

>> Range within which 50% of the values fall: 20 to 40

>> Range of the top quartile: 40 to 50

>> Range of the bottom quartile: 10 to 20

Present the following problem to your child:

Jimmy's class kept track of the number of books each child read during the past week:

Books Read

1

2

2

3

3

3

4

4

5

5

5

6

Using the following histogram, plot the frequencies of the number of books read.

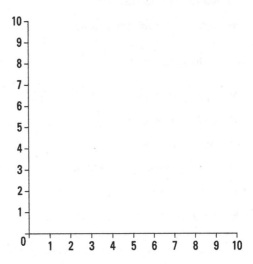

They should get:

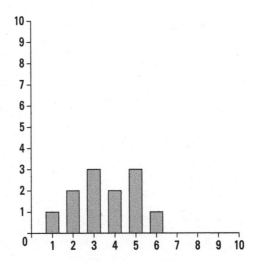

Chapter **12**

Making Sense of Ratios

R*atios.* Most of us recognize the term, but explaining them may be a different matter. You might remember, for example, a class in high school that had a high ratio of boys to girls, or vice versa. But if I told you the ratio of dogs to cats is 3:5, you might not recall how to read the ratio format.

Relax! This chapter shows you how to explain ratios to your child. You learn that a ratio lets you compare two things, such as the number of days off to the number of days worked, afternoons with naps to afternoons without, and the number of home team fans to visiting fans at a sporting event.

TIP

If you cook, you may find this chapter helpful. If a recipe for 10 cookies asks for you to use two eggs (that would be a wacky cookie recipe, but stick with me here), you can use proportional ratios to quickly determine how many eggs you would need to make 100 cookies.

Understanding Ratios

In this section, I provide you with what you need to know to teach your child about ratios. A *ratio* is a value that lets you compare two things. You might use a ratio to compare the number of students who play soccer to the number who play basketball, the cost of gas to the cost of wine, or the amount of time you spend worrying about ratios to the amount of time it will actually take you to teach your child to understand them!

Say to your child: "A ratio is a value that lets you compare two things. Assume, for example, that I have the following coins in my pocket."

Explain to your child: "You can use ratios to compare the number of coins. For example, I have 2 dimes and 3 pennies. I can write the ratio in two ways: using a fraction or separating the values with a colon. So, my ratio of dimes to pennies looks like this."

$$\frac{2}{3} \text{ or } 2{:}3$$

Tell them: "In a similar way, the ratio of my pennies to nickels is as follows."

$$\frac{3}{1} \text{ or } 3{:}1$$

Ask your child: "How would you represent the ratio of nickels to dimes?"

Your child should get:

$$\frac{1}{2} \text{ or } 1{:}2$$

TIP

If they do not, help them count the number of coins and tell them that the ratio's order (such as nickels to dimes) tells them which number is on top in the fraction (the numerator), or which comes first if you use the colon format.

Say to your child: "For example, if I ask, what is the ratio of dimes to pennies, you put the number of dimes on the top of the fraction or specify the number of dimes first, to the left of the colon."

$\frac{2}{3}$ or 2:3

Explain to them: "However, if I ask you the ratio of pennies to dimes, you put the number of pennies on top (or first)."

$\frac{3}{2}$ or 3:2

Ask your child to consider the following example:

Debbie's class has 10 boys and 12 girls.

Ask your child to provide the ratio of boys to girls. They should get:

$\frac{10}{12}$ or 10:12

Say to your child: "The ratio of birds to mice in the pet store is $\frac{5}{10}$ or 5:10. How many birds does the pet store have?"

They should say 5. If they don't, remind them that birds are the top or first number in the ratio, and mice are the bottom or the number to the right of the colon.

Ask your child to consider the following:

A ranch has 10 cows and 8 horses.

Ask them to represent the ratio of cows (10) to horses (8). They should get:

$\frac{10}{8}$ or 10:8

If they don't, remind them that the question is asking for the ratio of cows to horses. That means the number of cows is on the top of the fraction and appears before the colon.

Reducing Ratios (Equivalent Ratios)

In Chapter 6, your child reviewed the process for reducing fractions, such as:

$$\frac{4}{8} = \frac{1}{2} \qquad \frac{6}{10} = \frac{3}{5} \qquad \frac{8}{12} = \frac{2}{3}$$

In a similar way, when your child works with ratios, there may be times when they calculate ratios such as 10:5 or 4:8. In such cases, your child can reduce the ratios into equivalent ratios:

$$10:5 = 2:1 \qquad 4:8 = \frac{1}{2}$$

In this section, you teach your child to reduce ratios.

Say to your child: "You have learned that when you work with fractions, there are times when you must reduce the fractions to put them into a proper form."

$$\frac{5}{25} = \frac{1}{5} \qquad \frac{4}{10} = \frac{2}{5} \qquad \frac{9}{12} = \frac{3}{4} \qquad \frac{5}{10} = \frac{1}{2}$$

Explain to your child: "You have learned to represent ratios using fractions. When you write ratios, you should reduce them, just as you would any fraction."

Consider the following ratios:

$$5:25 = 1:5 \qquad 4:10 = 2:5 \qquad 9:12 = 3:4 \qquad 5:10 = 1:2$$

TIP

Say to your child: "You may find it easier to reduce ratios by representing the ratios as fractions and then reducing the fractions."

Ask your child to reduce the following ratios:

$$5:15 \qquad 4:16 \qquad 20:50 \qquad 7:49 \qquad 27:81$$

Your child should get:

$$1:3 \qquad 1:4 \qquad 2:5 \qquad 1:7 \qquad 1:3$$

If your child does not, review the following math with them:

$$\frac{5}{15} \div \frac{5}{5} = \frac{1}{3} \qquad \frac{4}{16} \div \frac{4}{4} = \frac{1}{4} \qquad \frac{20}{50} \div \frac{10}{10} = \frac{2}{5}$$
$$\qquad 1:3 \qquad\qquad\qquad 1:4 \qquad\qquad\qquad 2:5$$

$$\frac{7}{49} \div \frac{7}{7} = \frac{1}{7} \qquad \frac{27}{81} \div \frac{9}{9} = \frac{3}{9}$$
$$\qquad 1:7 \qquad\qquad\qquad 3:9$$

FIND ONLINE

This book's companion website at www.dummies.com/go/teachingyourkidsnew math6–8fd provides a worksheet (the first 3 rows of which are shown in Figure 12-1) that your child can use to reduce ratios. Download and print the worksheet. Help your child solve the first few and then ask them to complete the rest.

4:2	6:3	10:5	6:12
12:4	5:5	4:8	6:8
15:25	30:40	5:10	20:10

FIGURE 12-1:
A ratio worksheet.

Understanding Scale

In the previous section, your child learned to create simple ratios. In this section, they use ratios to understand the concept of scale.

Say to your child: "You have learned that ratios allow you to compare two objects. When you examine the blueprints for a building or a map of a city, you often see ratios used to specify the drawing's scale. Using the scale value, you can better compare the size of the drawing to the real-life equivalent."

Present the following map to your child:

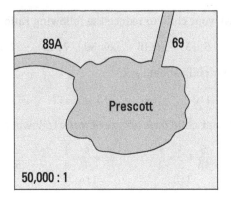

Tell your child: "This map represents the city of Prescott, Arizona. At the bottom of the map, you can see the scale of 50,000:1. The scale tells you that the city of Prescott is 50,000 times bigger than the map."

TIP

If you have a map or atlas available, look on the map to see if you can find a scale that you can show and discuss with your child. The U.S. Geological Survey has an excellent article on map scale at https://pubs.usgs.gov/unnumbered/70039582/report.pdf.

Understanding Proportions

This chapter has taught you how to teach your child to understand ratios. In this section, you examine a related topic, proportions.

Say to your child: "You have learned to use ratios to compare two things. You say that two values are in 'proportion' when a change in one of the values causes a related change in the other."

Ask your child to consider the following example:

Billy starts a business mowing lawns. Billy can mow 3 lawns a day.

Ask your child to represent the number of lawns mowed to the number of workers. They should get:

$\frac{3}{1}$ or 3:1

Tell your child: "Billy decides to hire his friend Shawna, who can also mow 3 lawns a day."

Say to your child: "Because Billy and Shawna can each mow the same number of lawns, you can create a proportion that tells you the total number of lawns the two can mow."

Present the following scenario to your child:

> Johnny wants to make cookies for the bake sale. The recipe calls for 2 eggs in order to make 12 cookies.

Ask your child to calculate the ratio of eggs to cookies. They should get:

$\frac{2}{12}$ or 2:12

Tell your child that Johnny wants to make 36 cookies.

Say to your child: "Using proportional ratios, you can determine how many eggs Johnny needs."

1. **Write the two ratios side by side as shown here:**

   $\frac{2}{12} = \frac{?}{36}$

   Since you don't know the number of eggs Johnny needs, you can represent them using a question mark.

2. **To solve for the number of eggs, have your child multiply 2 × 36:**

   $$\begin{array}{r} 36 \\ \times\, 2 \\ \hline 72 \end{array}$$

3. **Have your child divide 72 by 12:**

   $$\begin{array}{r} 6 \\ 12\overline{)\,72} \\ -72 \\ \hline 0 \end{array}$$

4. **Explain to your child that Johnny needs 6 eggs to make 36 cookies.**

Present the following scenario to your child:

> Sarah bought 1 bag of carrots to feed 4 horses.

Ask your child to calculate the ratio of carrot bags to horses. They should get:

$$\frac{1}{4} \text{ or } 1:4$$

Tell your child: "Sarah wants to feed carrots to 20 horses. How many bags of carrots does Sarah need?"

1. **Ask your child to write the ratios side by side:**

$$\frac{1}{4} = \frac{?}{20}$$

2. **Have your child multiply 1 × 20:**

$$\begin{array}{r} 20 \\ \times 1 \\ \hline 20 \end{array}$$

3. **Have your child divide 20 by 4:**

$$\begin{array}{r} 5 \\ 4\overline{)\ 20} \\ -20 \\ \hline 0 \end{array}$$

4. **Explain to your child that Sarah needs 5 bags of carrots to feed 20 horses.**

Present the following scenario to your child:

> Javier wants to start a shipping company to compete with Amazon. Using his bike, Javier can deliver 8 packages per day.

Ask your child to calculate the ratio of packages to workers. They should get:

$$\frac{8}{1} \text{ or } 8:1$$

Explain to your child: "On his first day, Javier has 48 packages to deliver. Javier decides to hire his friends to help. Assuming each of his friends can deliver 8 packages a day, how many friends does Javier need to help him?"

1. **Ask your child to write the ratios side by side:**

$$\frac{8}{1} = \frac{48}{?}$$

2. **Have your child multiply 1 × 48.**

$$\begin{array}{r} 48 \\ \underline{\times 1} \\ 48 \end{array}$$

3. **Have your child divide 48 by 8:**

$$\begin{array}{r} 6 \\ 8\overline{)\,48} \\ \underline{-48} \\ 0 \end{array}$$

4. **Explain to your child that Javier needs 6 friends to deliver the 48 packages.**

Working Out Word Problems with Ratios

You may be asking yourself, "What is the ratio of chapters with word problems to chapters without them?" That's a great question — ask your child to help!

Present the following word problem to your child:

> Anita wants to go to the pet store to look at puppies and kittens. The pet store has 3 puppies and 6 kittens. What is the ratio of kittens to puppies?

Your child should get:

$\frac{6}{3}$ or 6:3

TIP

Remind your child that when they work with ratios, they may get a ratio such as $\frac{6}{3}$, which, like a fraction, they can reduce to $\frac{2}{1}$ or 2:1.

Present the following word problem to your child:

> While eating lunch at a fast-food restaurant, Angela noticed that some customers bought French fries, while some did not. The ratio of customers who ate French fries to those who did not was $\frac{12}{5}$ or 12:5. How many customers bought French fries?

Your child should get 12.

Present the following word problem to your child:

> Mary has the goal of reading 30 books. She can read 1 book every 2 days. How many days will it take Mary to reach her goal?

1. **Have your child calculate the ratio of books read to days:**

   $\frac{1}{2}$ or 1:2

2. **Have your child write the two ratios side by side:**

   $\frac{1}{2} = \frac{30}{?}$

3. **Ask your child to multiply 2×30:**

   $$\begin{array}{r} 30 \\ \times 2 \\ \hline 60 \end{array}$$

4. **Ask your child to divide 60 by 1:**

   $$\begin{array}{r} 60 \\ 1{\overline{)}\,60} \\ -60 \\ \hline 0 \end{array}$$

5. **Tell your child that Mary will need 60 days to read 30 books.**

IN THIS CHAPTER

» Raising values to a power (exponent)

» Working with exponents of 0 and 1

» Considering negative exponents

» Solving square roots

» Multiplying and dividing values with exponents

Chapter **13**

Working with Exponents

E xponents — just the term alone sounds hard! As you may remember, an exponent is the power to which you raise another number, such as the 3 in $2^3 = 8$. In this chapter, you introduce your child to exponents such as the following:

$$3^2 = 3 \times 3 = 9 \qquad 5^2 = 5 \times 5 = 25 \qquad 6^2 = 6 \times 6 = 36$$

You may remember that raising a number to the power of 2 is called *squaring* the number. This chapter also introduces finding the *square root*, which is the opposite of finding the square:

$$\sqrt{9} = 3 \qquad \sqrt{25} = 5 \qquad \sqrt{36} = 6$$

Your child also works with the exponent values of 0 and 1, the results of which you may not remember!

$$3^1 = 3 \qquad 2^1 = 2 \qquad 3^0 = 1 \qquad 2^0 = 1$$

Then, your child works with negative exponents, such as 2^{-1}. And finally, they find out how to multiply and divide values with exponents.

$$2^{-1} = \frac{1}{2} \qquad 2^{-2} = \frac{1}{2^2} = \frac{1}{4} \qquad 2^{-3} = \frac{1}{2^3} = \frac{1}{8}$$

So, grab a power drink from the fridge and get ready to work with exponents (aka "powers").

Understanding Exponents

As stated in the chapter introduction, an *exponent* is the value (or power) to which an expression is raised, such as 5^2. In this section, you teach your child to understand and use exponents.

Say to your child: "When you perform mathematical operations, there will be times when you must multiply a number times itself."

$$3 \times 3 \quad 4 \times 4 \quad 5 \times 5 \quad 6 \times 6$$

Tell them: "In such cases, you can represent the multiplication operation by writing a 2 above the number you are multiplying."

$$3^2 \quad 4^2 \quad 5^2 \quad 6^2$$

Explain to your child: "The 2 in each of these expressions is called an *exponent*. You will also hear it called the *power*, as in 3 raised to the power of 2, or 4 raised to the power of 2. In addition, when the exponent is equal to 2, as in these examples, you will hear the exponent described as *squared*, as in 3 squared, 4 squared, and so on."

Ask your child to rewrite the following expressions using the exponent of 2:

$$1 \times 1 \quad 7 \times 7 \quad 8 \times 8 \quad 9 \times 9$$

They should get:

$$1^2 \quad 7^2 \quad 8^2 \quad 9^2$$

If they do not, show them the previous examples.

Present the following expressions to your child:

$$3 \times 3 \times 3 \quad 4 \times 4 \times 4 \quad 5 \times 5 \times 5 \quad 7 \times 7 \times 7$$

Say to your child: "In a similar way, you can represent these expressions using exponents. In this case, you use the power of 3."

$$3^3 \quad 4^3 \quad 5^3 \quad 7^3$$

Say to them: "When you raise a value to the exponent of 3, you say the expression has been *cubed*."

Ask your child to rewrite the following expressions using an exponent (power) of 3:

$$6 \times 6 \times 6 \quad 2 \times 2 \times 2 \quad 8 \times 8 \times 8 \quad 10 \times 10 \times 10$$

Your child should get:

6^3 $\qquad$ 2^3 $\qquad$ 8^3 $\qquad$ 10^3

REMEMBER

Explain to your child: "There is no limit on the value you can use for an exponent (power)."

Ask your child to rewrite the following expressions in their expanded form, such as $3^3 = 3 \times 3 \times 3$:

2^4 $\qquad$ 3^5 $\qquad$ 6^6 $\qquad$ 7^2

Your child should get:

$2 \times 2 \times 2 \times 2$ $\qquad$ $3 \times 3 \times 3 \times 3 \times 3$ $\qquad$ $6 \times 6 \times 6 \times 6 \times 6 \times 6$ $\qquad$ 7×7

TIP

If they do not, explain that the exponent tells you how many times you should multiply the number times itself.

FIND
ONLINE

This book's companion website at www.dummies.com/go/teachingyourkidsnew math6-8fd contains a worksheet (the first few rows of which are shown in Figure 13-1) that they can use to rewrite expressions using exponential form. Download and print the worksheet. Help your child solve the first few problems and then ask them to complete the rest.

2×2 $\qquad$ 3×3 $\qquad$ 4×4 $\qquad$ 5×5 $\qquad$ 6×6

FIGURE 13-1:
A worksheet
for writing
expressions in
exponential form.
$4 \times 4 \times 4$ $\qquad$ $3 \times 3 \times 3$ $\qquad$ $5 \times 5 \times 5$ $\qquad$ 7×7 $\qquad$ $6 \times 6 \times 6$

FIND
ONLINE

This companion website also contains a worksheet that lets your child practice rewriting expressions that use exponents in their expanded form. You can see the first few rows of this worksheet in Figure 13-2. Download and print the worksheet. Help your child solve the first few problems and then ask them to solve the rest.

FIGURE 13-2:
A worksheet
to rewrite
expressions
that use
exponents in
their expanded
form.
2^2 $\qquad$ 3^2 $\qquad$ 4^2 $\qquad$ 5^2 $\qquad$ 6^2

4^3 $\qquad$ 3^3 $\qquad$ 5^3 $\qquad$ 7^2 $\qquad$ 6^3

Explain to your child: "When you work with expressions that use exponents, you learn that expressions are not limited to numbers only." Ask them to consider the following expression:

$$(3+5)^2$$

Explain to your child: "In this case, you can rewrite the expression as follows."

$$(3+5)^2 =$$
$$(3+5)\times(3+5) =$$
$$8\times 8 = 64$$

Understanding the exponent of 1

Values can also have an exponent of 1, such as:

$$2^1 \qquad 3^1 \qquad 4^1 \qquad 6^1$$

REMEMBER

When you raise a value to an exponent of 1, the result is that value:

$$2^1 = 2 \qquad 3^1 = 3 \qquad 4^1 = 4 \qquad 6^1 = 6$$

Say to your child: "In the previous section, you used exponent values such as 2, 3, 4, 5, and so on. You can also use an exponent value of 1."

$$5^1 \qquad 6^1 \qquad 7^1 \qquad 8^1$$

Explain to your child: "When you raise a value to the power of 1, the result is that value."

$$5^1 = 5 \qquad 6^1 = 6 \qquad 7^1 = 7 \qquad 8^1 = 8$$

Understanding the exponent of 0

Values can also have an exponent of 0.

$$2^0 \qquad 3^0 \qquad 4^0 \qquad 6^0$$

REMEMBER

When you raise a value to an exponent of 0, the result is always 1:

$$2^0 = 1 \qquad 3^0 = 1 \qquad 4^0 = 1 \qquad 6^0 = 1$$

Say to your child: "In the previous section, you used an exponent of 1. It turns out that you can also use an exponent value of 0."

$$5^0 \qquad 6^0 \qquad 7^0 \qquad 8^0$$

WHY A VALUE RAISED TO 0 IS 1

Earlier in the chapter, you learn that any value you raise to the power to 0 is 1. At first glance, it may seem more logical for any number raised to the power of 0 to be 0. Why is it 1?

To understand why this is the case, consider the following:

$$2^4 \div 2^4 = 2^0$$

If you were to use the actual numbers ($2^4 = 16$), you would get:

$$16 \div 16 = 1$$

Therefore, $2^0 = 1$. Clear as mud?

Explain to your child: "When you raise a value to the power of 0, the result is always 1."

$$5^0 = 1 \qquad 6^0 = 1 \qquad 7^0 = 1 \qquad 8^0 = 1$$

Practicing Values Raised to the Power of 2

Your child will often encounter values from 1 through 10 raised to the power of 2 (squared). They should also review values raised to the power of 0 and 1. To help them quickly recognize the results of these expressions, use 3x5 index cards to create the following flashcards:

1^2	2^2	3^2	4^2	5^2	6^2	7^2	8^2	9^2	10^2
1^0	2^0	3^0	4^0	5^0	6^0	7^0	8^0	9^0	10^0
1^1	2^1	3^1	4^1	5^1	6^1	7^1	8^1	9^1	10^1

TIP

Practice the flashcards on a regular basis until your child masters them. Do not move on to negative exponents until they do.

FIND ONLINE

This book's companion website at www.dummies.com/go/teachingyourkidsnewmath6-8fd contains a worksheet (the first few rows of which are shown in Figure 13-3) that your child can use to practice values raised to the powers of 0, 1, and 2. Download and print the worksheet. Help your child solve the first few and then ask them to complete the rest.

2^2 3^2 4^2 5^2 6^2

4^1 3^0 5^1 7^2 6^0

What About Negative Exponents?

You have learned that your child can raise a value to a wide range of exponents (powers). It turns out that you can also raise a value to a negative exponent, such as:

2^{-2}

When you encounter a negative exponent, you rewrite it as follows:

$2^{-2} = \dfrac{1}{2^2} = \dfrac{1}{4}$

Consider the following:

2^{-3} 3^{-2} 4^{-2} 5^{-3}

$2^{-3} = \dfrac{1}{2^3} = \dfrac{1}{8}$ $\quad 3^{-2} = \dfrac{1}{3^2} = \dfrac{1}{9}$ $\quad 4^{-2} = \dfrac{1}{4^2} = \dfrac{1}{16}$ $\quad 5^{-3} = \dfrac{1}{5^3} = \dfrac{1}{125}$

Say to your child: "Just as you can raise an expression to a positive number, you can also raise a value to a negative power, such as 2^{-3}."

Tell them: "When you encounter a value raised to a negative power, you rewrite the expression as shown here."

$2^{-2} = \dfrac{1}{2^2} = \dfrac{1}{4}$

Ask your child to consider the following expression:

3^{-2}

Explain to them: "In this case, you rewrite the expression as follows."

$3^{-2} = \dfrac{1}{3^2} = \dfrac{1}{9}$

Ask your child to rewrite the following expressions:

4^{-2} 5^{-3} 6^{-1}

They should get:

4^{-2} 5^{-3} 6^{-1}

$4^{-2} = \dfrac{1}{4^2} = \dfrac{1}{16}$ $5^{-3} = \dfrac{1}{5^3} = \dfrac{1}{125}$ $6^{-1} = \dfrac{1}{6^1} = \dfrac{1}{6}$

FIND ONLINE

This book's companion website at www.dummies.com/go/teachingyourkidsnew math6-8fd contains a worksheet (the first few rows of which are shown in Figure 13-4) that your child can use to solve expressions raised to negative powers. Download and print the worksheet. Help your child solve the first few problems and then ask them to complete the rest.

2^{-2} 3^{-2} 4^{-2} 5^{-2} 6^{-2}

FIGURE 13-4:
A worksheet with
negative powers. 4^{-1} 3^{-1} 5^{-1} 7^{-2} 6^{-3}

Understanding the Square Root

In the previous sections, you taught your child how to square values:

$5^2 = 25$

$4^2 = 16$

$6^2 = 36$

In this section, you teach your child to use the *square-root* operator, which is essentially the opposite process of squaring a number. With the square root, you determine what value squared equals the number under the square-root symbol:

$\sqrt{25} = 5$ $\sqrt{16} = 4$ $\sqrt{36} = 6$

Say to your child: "In the previous section, you learned how to square values using an exponent of 2."

$2^2 = 4$

$3^2 = 9$

$5^2 = 25$

Tell them: "Now, you do just the opposite using the square-root symbol, which looks like this."

$$\sqrt{4} = 2 \qquad \sqrt{9} = 3 \qquad \sqrt{25} = 5$$

Explain to your child: "When you encounter the square-root operator, you ask yourself what value squared equals the number under the symbol." Ask your child to consider the following:

$$\sqrt{9} = 3 \qquad \sqrt{49} = 7 \qquad \sqrt{64} = 8$$

Ask your child to solve the following square roots:

$$\sqrt{81} \qquad \sqrt{100} \qquad \sqrt{36}$$

They should get:

$$\sqrt{81} = 9 \qquad \sqrt{100} = 10 \qquad \sqrt{36} = 6$$

If they don't, you should continue to practice the square-root flashcards until they have mastered them.

Using your 3x5 index cards, create the following square-root flashcards:

$\sqrt{4}$	$\sqrt{9}$	$\sqrt{16}$	$\sqrt{25}$	$\sqrt{36}$	$\sqrt{49}$	$\sqrt{64}$	$\sqrt{81}$	$\sqrt{100}$

TIP

Practice the flashcards on a regular basis until your child masters them. Do not move on to multiplying exponents until they do.

Multiplying Values with Exponents

When your child performs math operations with values that have exponents, they may encounter expressions similar to the following that multiply the same values that have exponents:

$$5^2 \times 5^2 = \qquad 4^3 \times 4^4 =$$

In such cases, you add the exponents as shown here:

$$5^2 \times 5^2 = 5^4 \qquad 4^3 \times 4^4 = 4^7$$

In this section, you teach your child to perform these processes.

Say to your child: "As you work with values that have exponents, there will be times when you must multiply the values, as in the following examples."

$$6^2 \times 6^2 =$$
$$3^3 \times 3^2 =$$

Explain to your child: "When the values are the same, as they are here, to perform the multiplication, you simply add the exponents."

$$6^2 \times 6^2 = 6^4$$
$$3^3 \times 3^2 = 3^5$$

If you solve the expressions, you can confirm they are equivalent:

$$6^2 \times 6^2 = 36 \times 36 = 1{,}296 \qquad 6^4 = 6 \times 6 \times 6 \times 6 = 1{,}296$$
$$3^3 \times 3^2 = 27 \times 9 = 243 \qquad 3^5 = 3 \times 3 \times 3 \times 3 \times 3 = 243$$

Ask your child to solve the following expressions:

$$2^2 \times 2^3 = \qquad 4^2 \times 4^2 = \qquad 6^2 \times 6^3 =$$

Your child should get:

$$2^2 \times 2^3 = 2^5 \qquad 4^2 \times 4^2 = 4^4 \qquad 6^2 \times 6^3 = 6^5$$

FIND ONLINE This book's companion website at www.dummies.com/go/teachingyourkidsnew math6-8fd contains a worksheet (the first few rows of which are shown in Figure 13-5) that your child can use to multiply values with exponents. Download and print the worksheet. Help your child solve the first few and then ask them to complete the rest.

$3^2 \times 3^2$ $2^3 \times 2^2$ $4^4 \times 4^2$ $7^2 \times 7^1$

FIGURE 13-5:
An exponent multiplication worksheet. $6^5 \times 6^2$ $7^3 \times 7^2$ $8^2 \times 8^2$ $9^3 \times 9^2$

Dividing Values with Exponents

As covered in the previous section, when you multiply values with exponents, you add the exponents. In a similar way, there will be times when your child must divide values with exponents:

$$5^4 \div 5^2 = \qquad 4^5 \div 4^2 =$$

REMEMBER

To do so, you simply subtract the exponents:

$$5^4 \div 5^2 = 5^2 \qquad 4^5 \div 4^2 = 4^3$$

In this section, you teach your child how to perform these processes.

Say to your child: "You have learned that when an expression multiplies the same value with exponents, you simply add the exponents as shown here."

$$5^2 \times 5^2 = 5^4 \qquad 3^3 \times 3^2 = 3^5$$

Tell them: "In a similar way, there are times when you must divide values with exponents, as in these examples."

$$3^4 \div 3^2 = \qquad 5^5 \div 5^2 =$$

Explain to your child: "In such cases, you simply subtract the exponents."

$$3^4 \div 3^2 = 3^2 \qquad 5^5 \div 5^2 = 5^3$$

Ask your child to solve the following expressions:

$$2^5 \div 2^3 = \qquad 4^4 \div 4^2 = \qquad 6^6 \div 6^3 =$$

Your child should get:

$$2^5 \div 2^3 = 2^2 \qquad 4^4 \div 4^2 = 4^2 \qquad 6^6 \div 6^3 = 6^3$$

FIND ONLINE

This book's companion website at www.dummies.com/go/teachingyourkidsnew math6-8fd contains a worksheet (the first few rows of which are shown in Figure 13-6) that your child can use to divide values with exponents. Download and print the worksheet. Help your child solve the first few and then ask them to complete the rest.

$3^3 \div 3^2$	$2^3 \div 2^2$	$4^4 \div 4^2$	$7^2 \div 7^1$
$6^5 \div 6^2$	$7^3 \div 7^2$	$8^2 \div 8^2$	$9^3 \div 9^2$

FIGURE 13-6:
An exponent division worksheet.

EXPONENTS IN REAL LIFE

Exponents are all around us:

- When you calculate the area of a room, you specify the area in square feet, such as 200 ft^2.

- When you calculate the area of a circle, you use the equation pi x radius x radius or pi x radius^2.

- Your computer likely has a terabyte disk, *tera* stands for 10^{12}.

- When you calculate a shape's volume, you specify the result in cubic dimensions, such as cubic feet.

- If you share your Instagram post with 10 people, who share it with 10 people, and those people share it with 10 people, 10^3 or 1,000 people will see your post.

IN THIS CHAPTER

» Using letters for variables and unknowns

» Performing the same operation to both sides of the equal sign

» Combining variables with the same terms

» Revisiting exponents

» Working with and visualizing inequalities

Chapter **14**

Getting Started with Algebra

lgebra. People don't remember why they don't like it; they just remember that it's often the last math they learned in school. Therefore, it often gets a bad rap.

Algebra is the branch of math that uses letters and symbols to represent things you don't know — called *unknowns.* Most people use algebra every day. You have $10 and gas costs $5 a gallon, so you know you can buy 2 gallons of gas. You have 3 friends for a barbecue and all 3 want 2 hot dogs, so you know you need 6 hot dogs. Using algebra, you have an unknown value for which you are trying to solve.

Algebra helps you solve such unknows. Since you don't know the value for which you are trying to solve, you represent it in equations using a letter, such as *x*:

$$x = 6 \div 2$$

So, get ready to dive into the unknown using algebra.

Algebra 101: Solving for x

Algebra uses letters and symbols to represent numbers within expressions. For example, consider the following expressions:

$$x = 3 + 1 \qquad x - 2 = 5 \qquad x^2 = 9$$

In this case, the letter x is called an unknown, meaning you don't know its value; therefore, you must solve for it. In the following sections, you teach your child how to solve such expressions.

Solving for x in simple expressions

Say to your child: "You are going to start learning algebra! It's a branch of math that uses letters and symbols to represent numbers. It's advanced math!"

Tell them: "In algebra, expressions use letters to represent unknown values. Your job is to solve for those values."

Ask your child to consider the following expressions:

$$x = 3 + 5 \qquad x = 2 \times 7 \qquad x = 25 - (3 \times 4)$$

Say to your child: "For these expressions, your job is to solve for x."

$$x = 3 + 5 \qquad x = 2 \times 7 \qquad x = 25 - (3 \times 4)$$
$$x = 8 \qquad\quad x = 14 \qquad\quad x = 25 - 12$$
$$x = 13$$

REMEMBER

Remind your child that for the third equation, $x = 25 - (3 \times 4)$, they must first solve the expression within the parentheses.

If your child does not get these results, help them perform the math.

FIND ONLINE

This book's companion website at www.dummies.com/go/teachingyourkids newmath6-8fd contains a worksheet (the first few rows of which are shown in Figure 14-1) that your child can use to solve expressions for x. Download and print the worksheet. Help your child with the first few and then ask them to solve the rest.

REMEMBER

Don't move on until your child has mastered solving simple expressions for x. Have your child practice until they master the process.

$$x = 5 + 3 \qquad x = 2 \times 3 \qquad x = 3 + 8 \qquad x = 25 \div 5$$

FIGURE 14-1:
A worksheet to
solve simple
expressions for *x*.

$$x = 4 + 2 \qquad x = 6 - 1 \qquad x = 9 - 3 \qquad x = 9 - 7$$

Solving for x in more complex expressions

Say to your child: "In the previous examples, you found the value of *x* by solving the expression on the opposite side of the equal sign."

$$x = 5 + 7$$
$$x = 12$$

Say to them: "Often, however, you will encounter expressions in the following form."

$$x - 5 = 7 \qquad x + 1 = 8 \qquad x + 2 = 5$$

Explain to your child: "When you encounter such expressions, you must perform operations to both sides of the equal sign with the goal of getting *x* by itself."

Ask your child to consider the following:

$$x - 5 = 7$$

Tell your child: "Your goal is to get *x* by itself on one side of the equal sign. In this case, to do that, you must add the value 5 to both sides of the equal sign."

$$x - 5 = 7$$
$$x - 5 + \mathbf{5} = 7 + \mathbf{5}$$

Say to your child: "In this case, by adding 5 to both sides of the expression, you get *x* by itself."

$$x = 7 + 5$$
$$x = 12$$

In the previous example, you added 5 to both sides of the expression.

$$x - 5 + 5 = 7 + 5$$

Here is the same operation, but with the addition written underneath:

$$x - 5 = 7$$
$$\underline{+5 \quad +5}$$
$$x = 12$$

Explain to your child: "In algebra, whatever operation you perform on one side of the equal sign, you must also do to the other side — that keeps the sides equal."

Ask your child to consider the following expression:

$$x + 1 = 8$$

Say to your child: "In this case, to get x by itself, you must subtract 1 from both sides."

$$x + 1 = 8$$
$$x + 1 - 1 = 8 - 1$$
$$x = 7$$

Ask your child to consider the following:

$$x + 2 = 5$$

Say to them: "To get x by itself, you must subtract 2 from both sides."

$$x + 2 = 5$$
$$x + 2 - 2 = 5 - 2$$
$$x = 3$$

Explain to your child: "After you solve for the value x, you can check your work by replacing the value of x back into the original expression."

$$x + 2 = 5$$
$$3 + 2 = 5$$
$$5 = 5$$

Ask your child to solve the following expressions:

$$x + 4 = 7 \qquad x + 5 = 12 \qquad x - 3 = 10$$

Your child should get:

$$x + 4 = 7 \qquad\qquad x + 5 = 12 \qquad\qquad x - 3 = 10$$
$$x + 4 - 4 = 7 - 4 \qquad x + 5 - 5 = 12 - 5 \qquad x - 3 + 3 = 10 + 3$$
$$x = 3 \qquad\qquad\qquad x = 7 \qquad\qquad\qquad x = 13$$

If your child does not, help them perform the math to get x by itself on one side of the equal sign.

This book's companion website at www.dummies.com/go/teachingyourkids newmath6-8fd contains a worksheet (the first few rows of which are shown in Figure 14-2) that your child can use to solve for x. Download and print the worksheet. Help your child solve the first few and then ask them to solve the rest.

$$x + 3 = 8 \qquad x + 1 = 5 \qquad x + 3 = 7 \qquad x + 3 = 5$$

FIGURE 14-2:
A worksheet to solve for *x* in more complex expressions.

$$x - 2 = 4 \qquad x - 3 = 9 \qquad x - 6 = 3 \qquad x - 2 = 7$$

Solving expressions that include multiples of x

Say to your child: "Sometimes you have expressions that include multiples of *x*."

$$2x = 10 \qquad 3x = 12 \qquad 5x = 25$$

Explain to them: "As before, your goal is to get *x* by itself on one side of the equal sign. In this case, you need to divide both equations by the same number."

Ask your child to consider the following expression:

$$2x = 10$$

Explain to them: "To get *x* by itself, you must divide both sides by 2."

$$2x = 10$$
$$\frac{2x}{2} = \frac{10}{2}$$
$$x = 5$$

Ask your child to consider the following:

$$3x = 12$$

Explain to your child: "To get *x* by itself, you must divide both sides by 3."

$$3x = 12$$
$$\frac{3x}{3} = \frac{12}{3}$$
$$x = 4$$

Finally, ask your child to consider the following:

$$5x = 25$$

Say to them: "To get *x* by itself, you must divide both sides by 5."

$$5x = 25$$
$$\frac{5x}{5} = \frac{25}{5}$$
$$x = 5$$

Tell your child: "As before, you can test your result (5) by substituting it for x in the original expression."

$$5x = 25$$
$$5 \times 5 = 25$$
$$25 = 25$$

FIND ONLINE

This book's companion website at www.dummies.com/go/teachingyourkids newmath6–8fd contains a worksheet (the first few rows of which are shown in Figure 14-3) that your child can use to solve for expressions that have multiples of x. Download and print the worksheet. Help your child solve the first few and then ask them to solve the rest.

$6x = 36$	$7x = 14$	$3x = 6$
$2x = 4$	$5x = 25$	$6x = 42$
$8x = 56$	$9x = 90$	$10x = 20$

FIGURE 14-3: A worksheet for solving for multiples of x.

Performing multiple operations to solve for x

Say to your child: "There may be times when you must perform several operations to get the value of x by itself."

$$2x + 1 = 11 \qquad 7x - 3 = 18 \qquad 5x + 5 = 25$$

Ask your child to consider the following expression:

$$2x + 1 = 11$$

1. **Have your child subtract 1 from both sides of the expression:**

$$2x + 1 - 1 = 11 - 1$$
$$2x = 10$$

2. **Have them divide both sides by 2:**

$$2x = 10$$
$$\frac{2x}{2} = \frac{10}{2}$$
$$x = 5$$

Have your child consider the following expression:

$$7x - 3 = 18$$

1. **Have your child add 3 to both sides of the expression:**

$$7x - 3 + 3 = 18 + 3$$
$$7x = 21$$

2. **Have them divide both sides of the expression by 7:**

$$7x = 21$$
$$\frac{7x}{7} = \frac{21}{7}$$
$$x = 3$$

Consider the following expression:

$$5x + 5 = 25$$

1. **Have your child subtract 5 from both sides of the expression:**

$$5x + 5 - 5 = 25 - 5$$
$$5x = 20$$

2. **Have them divide both sides of the expression by 5:**

$$5x = 20$$
$$\frac{5x}{5} = \frac{20}{5}$$
$$x = 4$$

Ask your child to solve the following:

$$2x + 7 = 9 \qquad 7x - 7 = 14 \qquad 25x + 5 = 30$$

Your child should get:

$$2x + 7 = 9 \qquad\qquad 7x - 7 = 14 \qquad\qquad 25x + 5 = 30$$
$$2x + 7 - 7 = 9 - 7 \qquad 7x - 7 + 7 = 14 + 7 \qquad 25x + 5 - 5 = 30 - 5$$
$$2x = 2 \qquad\qquad 7x = 21 \qquad\qquad 25x = 25$$
$$\frac{2x}{2} = \frac{2}{2} \qquad\qquad \frac{7x}{7} = \frac{21}{7} \qquad\qquad \frac{25x}{25} = \frac{25}{25}$$
$$x = 1 \qquad\qquad x = 3 \qquad\qquad x = 1$$

If your child does not, work through each step of these solutions, which gets *x* by itself on one side of the equal sign.

As before, your child should check their results by substituting the value for x in the original expressions:

$$2x+7=9 \qquad\qquad 7x-7=14 \qquad\qquad 25x+5=30$$
$$2\times1+7=9 \qquad\qquad 7\times3-7=14 \qquad\qquad 25\times1+5=30$$
$$2+7=9 \qquad\qquad 21-7=14 \qquad\qquad 25+5=30$$
$$9=9 \qquad\qquad 14=14 \qquad\qquad 30=30$$

FIND ONLINE

This book's companion website at www.dummies.com/go/teachingyourkids newmath6-8fd contains a worksheet (the first few rows of which are shown in Figure 14-4) that your child can use to solve for x. Download and print the worksheet. Help your child solve the first few and then ask them to solve the rest.

$$2x-2=6 \qquad\quad 2x-1=9 \qquad\quad 5x-2=8 \qquad\quad 5x-5=25$$

FIGURE 14-4:
A worksheet for
solving for x.

$$3x+3=4+2 \qquad 5x-0=6-1 \qquad 3x-3=9-3 \qquad 2x-1=9-4$$

Simplifying Expressions

When your child works with algebraic expressions, there will be times when they first need to organize the expression before they can solve it. This process is called *simplifying* the expression.

Ask your child to consider the following expression:

$$x+x=10$$

In this case, they must solve for x; however, there are two of them and their values must be the same. To simplify the expression, you rewrite it as:

$$2x=10$$

Ask your child to simplify the following expression:

$$x+x+x=15$$

They should get:

$$3x=15$$

If they do not, have them count and rewrite the number of x's.

Ask your child to simplify the following expressions:

$$2x + x = 18 \qquad\qquad 3x + 2x = 15$$

They should get:

$$3x = 18 \qquad\qquad 5x = 15$$

If they do not, have them count and rewrite the number of x's.

Revisiting Exponents with Unknowns

In Chapter 13, your child learned to work with exponents such as:

$$2^2 = 4 \qquad 5^2 = 25 \qquad 7^2 = 49$$

In this section, your child works with unknowns that have exponents:

$$x^2 = 25 \qquad x^2 = 49 \qquad x^2 = 64$$

Solving for x with exponents

Say to your child: "You have learned to work with numbers that have exponents, such as the following."

$$3^2 = 9 \qquad 4^2 = 16 \qquad 10^2 = 100$$

Explain to your child: "When you perform algebra, there are times when variables have exponents."

$$x^2 = 36 \qquad x^2 = 81 \qquad x^2 = 4$$

Say to your child: "To solve for x in such cases, you must take the square root of both sides of the equal sign.

Ask your child to consider the following:

$$x^2 = 81$$

Explain to them: "Again, you must take the square root of both sides."

$$x^2 = 81$$
$$\sqrt{x^2} = \sqrt{81}$$
$$x = \sqrt{81}$$
$$x = 9$$

Ask your child to solve the following:

$$x^2 = 144 \qquad x^2 = 1 \qquad x^2 = 49$$

They should get:

$$\begin{array}{ccc}
x^2 = 144 & x^2 = 1 & x^2 = 49 \\
\sqrt{x^2} = \sqrt{144} & \sqrt{x^2} = \sqrt{1} & \sqrt{x^2} = \sqrt{49} \\
x = \sqrt{144} & x = \sqrt{1} & x = \sqrt{49} \\
x = 12 & x = 1 & x = 7
\end{array}$$

FIND ONLINE

This book's companion website at www.dummies.com/go/teachingyourkids newmath6-8fd has a worksheet (the first few rows of which are shown in Figure 14-5) that your child can use to solve for x when exponents are used. Download and print the worksheet. Help your child solve the first few and then ask them to solve the rest.

$$x^2 = 4 \qquad\qquad x^2 = 9 \qquad\qquad x^2 = 25 \qquad\qquad x^2 = 16$$

FIGURE 14-5:
A worksheet to solve for x with exponents.

$$x^2 = 100 \qquad\qquad x^2 = 49 \qquad\qquad x^2 = 64 \qquad\qquad x^2 = 81$$

Solving for x with complex expressions

Ask your child to consider the following expression:

$$2x^2 + 4 = 12$$

Say to your child: "In this case, you must perform multiple steps to solve for x."

1. **Have your child subtract 4 from both sides:**

$$\begin{array}{c}
2x^2 + 4 = 12 \\
2x^2 + 4 - 4 = 12 - 4 \\
2x^2 = 8
\end{array}$$

2. **Have them divide both sides of the expression by 2:**

$$\begin{array}{c}
\dfrac{2x^2}{2} = \dfrac{8}{2} \\
x^2 = 4
\end{array}$$

3. **Have your child take the square root of both sides:**

$$x^2 = 4$$
$$\sqrt{x^2} = \sqrt{4}$$
$$x = \sqrt{4}$$
$$x = 2$$

Ask your child to consider the following expression:

$$3x^2 - 3 = 45$$

1. **Have your child add 3 to both sides:**

$$3x^2 - 3 = 45$$
$$3x^2 - 3 + 3 = 45 + 3$$
$$3x^2 = 48$$

2. **Have them divide both sides by 3:**

$$3x^2 = 48$$
$$\frac{3x^2}{3} = \frac{48}{3}$$
$$x^2 = 16$$

3. **Have your child take the square root of both sides:**

$$x^2 = 16$$
$$\sqrt{x^2} = \sqrt{16}$$
$$x = \sqrt{16}$$
$$x = 4$$

FIND ONLINE

This book's companion website at www.dummies.com/go/teachingyourkids newmath6-8fd contains a worksheet (the first few rows of which are shown in Figure 14-6) that your child can use to solve for x. Download and print the worksheet. Help your child solve the first few and then ask them to complete the rest.

$2x^2 = 8$	$3x^2 = 27$	$4x^2 = 16$	$2x^2 = 18$
$5x^2 = 125$	$2x^2 = 18$	$2x^2 = 72$	$3x^2 = 27$

FIGURE 14-6: A worksheet for solving for x with complex expressions.

Solving Inequalities

In the previous sections, your child learned to solve for x as an equality — meaning the expression used an equal sign:

$x = 5 + 1$

In this section, your child performs similar operations with inequalities — expressions that use $<, >, \geq$, and $\leq$. Here are a couple of examples:

$x < 7 + 1 \qquad x \geq 5 \times 2$

Say to your child: "In the previous section, you solved for the value of x. But there will be times when you encounter expressions that use inequalities, such as $>, <, \geq$, and $\leq$. Consider the following."

$x > 5 + 9 \qquad x + 1 < 8 \qquad x \geq 5 \times 7$

Tell your child: "You solve such expressions using the same steps that you just followed for equalities (expressions with equal signs)."

$$
\begin{array}{lll}
x > 5 + 9 & x + 1 < 8 & x \geq 5 \times 7 \\
x > 14 & x + 1 - 1 < 8 - 1 & x \geq 35 \\
& x < 7 &
\end{array}
$$

WHY YOU USE INEQUALITIES

Often, when you perform math, you don't need to solve for a specific value, but rather, a range of values.

Assume you are creating your monthly budget. You have $500 after you pay your bills, and you want to do something fun. That said, you want to have a few dollars left when you are done. So if you spend less than $500, you can accomplish both of your goals:

Money spent on fun < $500

As long as you spend less than $500, you can have fun and money left over.

Illustrating Inequalities

Your child just learned to solve expressions with inequalities. Often, your child will need to represent their solutions visually or to explain such illustrations. In this section, they will do just that.

Representing an inequality visually

As your child works with inequalities, such as $x > 6$, they may be asked to represent the inequality visually, as shown here:

In this section, you teach your child to represent such inequalities.

Say to your child: "When you work with inequalities, there may be times when you are asked to visually represent the inequality using a number line. Consider this expression."

$x > 5$

Tell your child: "Using a number line, you can represent that x is greater than 5 as follows."

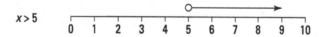

Explain to them: "The open circle and arrow above the number 5 in the line indicates 'greater than 5.'"

Ask your child to consider the following expression:

$x < 4$

Say to them: "To represent this inequality, you use the open circle and arrow as follows:"

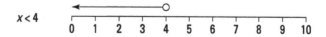

Explain to your child: "Again, in this case, the open circle and left-facing arrow above the 4 represent that x is less than 4."

Ask your child to consider the following expression:

$x \geq 3$

Explain to them: "In this case, because you are using ≥, you use a closed circle as shown here."

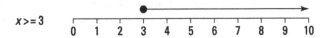

Ask your child to visually represent the following inequalities:

$x \geq 5$ $x < 7$ $x \leq 2$

They should get:

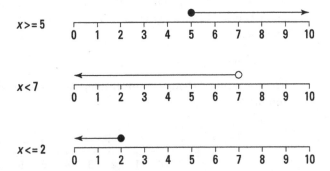

REMEMBER

If they don't, review the concept that they use an open circle for > and < and a closed circle for ≥ and ≤. Also, they use a right-facing arrow for values that are > and ≥ and a left-facing arrow for values that are < and ≤.

**FIND
ONLINE**

This book's companion website at www.dummies.com/go/teachingyourkids newmath6-8fd contains a worksheet (the first few rows of which are shown in Figure 14-7) that your child can use to represent inequalities. Download and print the worksheet. Help your child solve the first few and then ask them to solve the rest.

$$x > 5 + 3 \qquad\qquad x < 2 \times 3 \qquad\qquad x > 3 + 8 \qquad\qquad x < 25 \div 5$$

FIGURE 14-7:
A worksheet for
representing
inequalities.

$$x > 4 + 2 \qquad\qquad x < 6 - 1 \qquad\qquad x > 9 - 3 \qquad\qquad x < 9 - 7$$

Interpreting visualized inequalities

Say to your child: "In the previous section, you visualized inequalities. In this section, you do the opposite: you write the expression that is visualized."

Ask your child to consider the following:

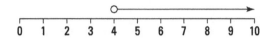

In this case, they should write:

$x > 4$

Ask your child to write the inequalities for the following visualizations:

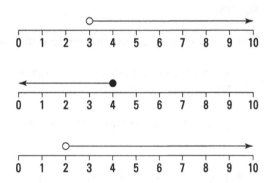

They should get:

$x > 3 \qquad\qquad x \leq 4 \qquad\qquad x > 2$

If they do not, remind them that an open circle is for < and > and a closed circle is for ≥ and ≤. Also, remind them that the right-facing arrow is for > and ≥ and the left-facing arrow is for < and ≤.

FIND ONLINE

This book's companion website at www.dummies.com/go/teachingyourkids newmath6–8fd contains a worksheet (the first few rows of which are shown in Figure 14-8) that your child can use to write the inequalities for visualizations. Download and print the worksheet. Help your child solve the first few and then ask them to solve the rest.

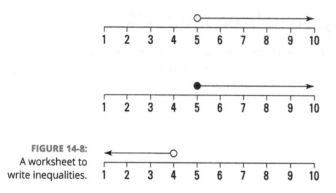

FIGURE 14-8:
A worksheet to
write inequalities.

Solving Algebra Word Problems

This section provides your child with practice solving word problems that use algebra.

Present the following word problem to your child:

> Billy, Mary and Ted need to sell 15 boxes of cookies. Billy sold 5 and Mary sold 3. How many must Ted sell?

Help your child construct the following expression:

$$15 = 5 + 3 + x$$

Have your child solve for x. They should get 8.

Present the following problem to your child:

> Juan's family is driving 1,500 miles to the ocean. On day 1, they drove 600 miles. On day 2, they drove 425 miles. How many miles must they drive on day 3?

Help your child construct the following problem:

$$1500 = 600 + 425 + x$$

Have your child solve for x. They should get 475.

Present the following problem:

> Billy and Sally have 12 books. Sally has twice as many books as Billy. How many books do each have?

In this case, you will represent the number of books that Billy has as x. Sally, therefore, has $2x$ books. Help your child construct the following problem:

$$12 = x + 2x$$

Help your child solve the problem:

$$12 = 3x$$

$$x = 4$$

If Billy has 4 books, Sally has $2x$ books, or $2 \times 4 = 8$ books.

Chapter **15**

Calculating and Applying Percentages

A 15 percent tip is good! A 30 percent score on a math test, not so good! Percentages, they're all around us. There are '25 percent off' signs in stores. There's a sales-tax percentage when you buy things, and your odds of having word problems at the end of a chapter are greater than 75 percent!

This chapter drills into percentages. You learn how to apply them (which is convenient in restaurants). You learn how much a big discount will save you. And, you learn how to determine what percentage one value is of another value.

So, get ready to master percentages. When you do well, give yourself a tip!

Getting Started with Percentages

A percentage is a fraction with 100 as the denominator (the bottom number) and written in decimal form. The common 15 percent tip, for example, is $\frac{15}{100}$. In other words, percentages are always expressed in terms of hundredths:

$$5\% = \frac{5}{100}$$

$$25\% = \frac{25}{100}$$

$$100\% = \frac{100}{100}$$

REMEMBER

In fact, if you want to impress your friends, you can tell them that *percent* means "out of one hundred."

In this section, you introduce your child to percentages.

Say to your child: "A percentage is a value, normally from 0 to 100, that you use to describe and compare things. For example, you might use percentage values for the following."

>> Providing a test score

>> Describing how much of a project you have completed

>> Calculating how much to tip the server at a restaurant

Say to your child: "A TV commercial states that 3 out of 4 dentists recommend a specific toothpaste. Assume you want to represent $\frac{3}{4}$ as a percentage."

1. Say to your child: "To start, you must change the fraction $\frac{3}{4}$ to its decimal form."

$$\frac{3}{4} = 4\overline{)3.00} \begin{array}{r} 0.75 \\ \hline \\ \underline{-28} \\ 20 \\ \underline{-20} \\ 0 \end{array}$$

2. Multiply the decimal value by 100 to produce the percentage:

$$0.75 \times 100 = 75\%$$

Say to your child: "To get started, assume a soccer team won 17 out of 25 games. You want to know the percentage of games the team won."

1. **Explain to your child: "To start, you can use the ratio of games won to games played, $\frac{17}{25}$."**

2. **Say to your child: "Next, you convert the fraction to its decimal form."**

$$
\begin{array}{r}
0.68 \\
25{\overline{\smash{\big)}\,17.00}} \\
\underline{-150} \\
200 \\
\underline{-200} \\
0
\end{array}
$$

3. **Say to them: "Finally, to convert the decimal value to a percentage, you multiply it by 100."**

$$
\begin{array}{r}
0.68 \\
\times\,100 \\
\hline
000 \\
0000 \\
6800 \\
\hline
68.00 \ \ \text{or} \ \ 68\%
\end{array}
$$

Ask your child to convert the following fractions to percentages:

$$\frac{1}{2} \qquad\qquad \frac{1}{4} \qquad\qquad \frac{1}{10}$$

They should get:

$$
\frac{1}{2} \ = \ 2{\overline{\smash{\big)}\,1.0}} \quad\ \ \
\begin{array}{r}
0.5 \\
\underline{-10} \\
0
\end{array}
\qquad = \quad
\begin{array}{r}
100 \\
\times\,0.5 \\
\hline
50\%
\end{array}
$$

$$
\frac{1}{4} \ = \ 4{\overline{\smash{\big)}\,1.00}}\quad
\begin{array}{r}
0.25 \\
\underline{-0.8} \\
20 \\
\underline{-20} \\
0
\end{array}
\qquad = \quad
\begin{array}{r}
100 \\
\times\,0.25 \\
\hline
25\%
\end{array}
$$

$$
\frac{1}{10} \ = \ 10{\overline{\smash{\big)}\,1.0}}\quad
\begin{array}{r}
0.1 \\
\underline{-10} \\
0
\end{array}
\qquad = \quad
\begin{array}{r}
100 \\
\times\,0.1 \\
\hline
10\%
\end{array}
$$

FIND ONLINE

This book's companion website at www.dummies.com/go/teachingyourkidsnew math6–8fd contains a worksheet (the first few rows of which are shown in Figure 15-1) that your child can use to practice representing fractions as percentages. Download and print the worksheet. Help your child solve the first few and then ask them to complete the rest.

$$\frac{1}{2} \qquad\qquad \frac{1}{4} \qquad\qquad \frac{1}{3} \qquad\qquad \frac{1}{5}$$

FIGURE 15-1:
A worksheet to
convert fractions
to percentages.

$$\frac{3}{4} \qquad\qquad \frac{4}{8} \qquad\qquad \frac{3}{5} \qquad\qquad \frac{2}{3}$$

How Much Should I Tip?

It feels like you're expected to tip nearly everyone these days. The challenge for many people is calculating a tip amount on the fly. In this section, you teach your child to do just that.

Say to your child: "When you eat at a restaurant, get a haircut, or even get your lawn mowed, you often leave a tip to recognize an individual's good service."

Say to them, "Many people leave a 15 percent tip for these kinds of services. Now you will learn how to calculate the dollar amount that corresponds to 15 percent."

Now give your child the following example to work on: "Assume your dinner bill is $30 and you want to leave a 15 percent tip."

1. Tell your child: "Start by representing 15% in its decimal form."

   $15\% = 0.15$

2. Explain to them: "You now multiply the total, which in this case is 30, by the decimal amount."

$$\begin{array}{r} 30 \\ \times\,0.15 \\ \hline 150 \\ 300 \\ \hline 4.50 \end{array}$$

3. Tell your child: "In this case, a 15 percent tip on $30 is $4.50."

Ask your child to consider the following scenario:

Jim is very happy with his haircut and wants to leave a 20 percent tip. The cost of the haircut is $25. How much money should Jim leave as a tip?

1. **Ask your child to convert 20 percent to its decimal form:**

$20\% = 0.20$

2. **Have your child multiply the total, in this case $25, by the decimal amount:**

$$
\begin{array}{r}
25 \\
\times\ 0.20 \\
\hline
00 \\
\underline{500}
\end{array}
$$

3. **Say to your child: "In this case, the tip amount is $5.00."**

Tell your child: "When you need to calculate a tip amount, you often ask yourself a question such as the following:"

What is 15% of 35?

Explain to your child: "In Chapter 13, you practiced solving problems for the variable x, an unknown value. If, within the previous question, you substitute x for the word 'What,' and substitute an equal sign for the word 'is' and a multiplication sign for the word 'of,' the question becomes the following."

What is 15% of 35?

$x = 15\% \times 35$

When you solve for x, you find the tip amount:

1. **Ask your child to convert 15 percent to its decimal form:**

$15\% = 0.15$

2. **Have your child multiply the total, in this case 35, by the decimal amount:**

$$
\begin{array}{r}
35 \\
\times\ 0.15 \\
\hline
175 \\
\underline{350} \\
5.25
\end{array}
$$

3. **Say to your child: "In this case, 15% of 35 is 5.25."**

Ask your child to solve the following tip amounts:

$40 and 20% tip $90 and 15% tip

They should get:

$$\begin{array}{r} 40 \\ \times\,0.20 \\ \hline 0 \\ 800 \\ \hline 8.00 \end{array} \qquad\qquad \begin{array}{r} 90 \\ \times\,0.15 \\ \hline 450 \\ 900 \\ \hline 13.50 \end{array}$$

Ask your child to consider the following scenario:

Bill wants to be a scuba diver. To do so, he must pass a test with a score of 70% or higher. There are 40 questions on the test. How many questions must Bill get right?

1. **Tell your child to start with the question:**

   What is 70% of 40?

2. **Replace the word "What" with x, "is" with an equal sign, and "of" with a multiplication sign:**

   $x = 70\% \times 40$

3. **Replace 70% with its decimal form:**

   $x = 0.70 \times 40$

4. **Solve for x:**

   $$\begin{array}{r} 40 \\ \times\,0.70 \\ \hline 0 \\ 2800 \\ \hline 28.00 \end{array}$$

5. **Say to your child that Bill must get 28 questions right.**

How Much Will I Save?

It's hard to walk through a retail store that does not have signs with discounts such as 20 percent off! In this section, you teach your child how to determine the actual savings.

Say to your child: "Assume you are shopping for jeans that cost $40. A sign above the jeans says '20% off.' After the discount, how much would you pay for the jeans?"

1. Ask your child: "If the discount on the jeans is 20 percent, what percentage of the jeans' price will you pay?"

They should say 80 percent. If they don't, help them subtract the 20 percent from 100 percent.

2. Say to your child: "The question, then, becomes this."

What is 80% of 40?

3. Replace the word "What" with x, "is" with an equal sign, and "of" with a multiplication sign:

$x = 80\% \times 40$

4. Replace 80% with its decimal form:

$x = 0.80 \times 40$

5. Solve for x:

$$\begin{array}{r} 40 \\ \times\ 0.80 \\ \hline 0 \\ 3200 \\ \hline 32.00 \end{array}$$

6. Explain to your child that with the discount, the jeans would cost $32.

Ask your child to consider the following scenario:

Sara wants to buy tacos for lunch. Tacos normally cost $2.50. However, today is "Taco Tuesday," and the restaurant discounts tacos by 30%. How much will Sara spend on each taco?

1. Ask your child: "If the discount on the tacos is 30%, what percentage of the price will Sara pay for each taco?"

They should say 70%. If they don't, help them subtract 30% from 100%.

2. Say to your child: "The question, then, becomes this."

What is 70% of 2.50?

3. Replace the word "What" with x, "is" with an equal sign, and "of" with a multiplication sign:

$x = 70\% \times 2.50$

4. Replace 70% with its decimal form:

$x = 0.70 \times 2.50$

5. **Solve for *x*:**

$$\begin{array}{r} 2.50 \\ \times\ 0.70 \\ \hline 0 \\ 17500 \\ \hline 1.75 \end{array}$$

6. **Say to your child that with the discount, a taco will cost $1.75.**

Determining What Percent One Value is of Another

In the previous sections, you taught your child how to multiply a percent times an amount to get a result. In this section, you teach your child how to determine the percent one value is of another, such as:

7 is what percent of 12?

To determine the percent, you divide 7 by 12:

$$7 \div 12 = 58\%$$

Say to your child: "In the previous section, you learned how to multiply a percent times an amount to calculate a result — such as a 15 percent tip amount."

Explain to your child: "Now you will learn how to determine what percent one value is of another, such as 8 is what percent of 20?"

Say to them: "To solve this problem, you now solve for the percent value."

1. **Say to your child: "To calculate what percent 8 is of 20, start with the fraction $\frac{8}{20}$."**

2. **Ask your child to perform the division, dividing 8 by 20:**

$$\frac{8}{20} = 20\overline{)\begin{array}{c} 0.4 \\ 80 \\ \underline{-80} \\ 0 \end{array}}$$

3. **Multiply the result, 0.40, by 100 to get the percentage:**

$$0.40 \times 100 = 40\%$$

4. **Say to your child: "8 is 40% of 20."**

Say to your child, "You may be wondering how I knew to start with the fraction $\frac{8}{20}$. Consider the original question."

8 is what percent of 20?

Explain to them: "As you did before, replace the word 'is' with an equal sign, the 'what' with x, and the 'of' with a multiplication sign."

8 is what percent of 20?

$8 = x\% \times 20$

Say to your child: "To solve for x, you divide both sides of the equal sign by 20."

$\frac{8}{20} = x$

Ask your child to solve the following:

27 is what percent of 30?

1. **Help your child rewrite the question into an expression:**

   $27 = x\% \times 30$

2. **Solve for x:**

   $\frac{27}{30} = x$

   $$\frac{27}{30} = 30\overline{)\begin{array}{r} 0.9 \\ 27.0 \\ -270 \\ \hline 0 \end{array}} = 0.9$$

3. **Ask your child to multiply the result, 0.9, by 100 to produce the percentage:**

   $0.9 \times 100 = 90\%$

FIND ONLINE

This book's companion website at www.dummies.com/go/teachingyourkidsnew math6-8fd contains a worksheet (the first few rows of which are shown in Figure 15-2) that your child can use to determine the percent one value is of another. Download and print the worksheet. Help your child solve the first few and then ask them to complete the rest.

3 is what percent of 4?

1 is what percent of 2?

3 is what percent of 5?

4 is what percent of 8?

FIGURE 15-2: A worksheet for solving the percent one value is of another.

Solving Problems Like "10 is 15% of What Number?"

Your child has learned to calculate the percentage of a number:

What is 15 percent of 90?

Also, your child has learned to calculate what percent of a value is a number:

7 is what percent of 13?

In this section, your child will be able to answer questions such as this:

13 is 12 percent of what number?

Say to your child: "There may be times when you know a number, such as 15, is some percentage, say 30 percent, of another number."

15 is 30% of what?

Say to your child: "To solve this problem, you need to create an expression and solve for x."

1. **Rewrite the question as an expression:**

   $15 = 30\% \times x$

2. **Solve for *x*:**

   $$15 = 30\% \times x$$
   $$\frac{15}{0.3} = x$$
   $$50 = x$$

3. **Say to your child: "15 is 30% of 50."**

Ask your child to consider the following:

23 is 12% of what?

1. **Rewrite the question as an expression:**

$$23 = 12\% \times x$$

2. **Solve for x:**

$$23 = 12\% \times x$$
$$\frac{23}{0.12} = x$$
$$192 = x$$

3. **Say to your child: "With a little rounding, 23 is 12% of 192."**

Working through Word Problems with Percentages

More than 75 percent of this book's chapters have word problems, and this is one of them!

Present the following word problem to your child:

Javier's family had a great dinner and wants to leave a 20 percent tip. The cost of the dinner was $80. What tip amount should they leave?

1. **Write the question asked:**

What is 20% of 80?

2. **Rewrite the question as an expression:**

$$x = 20\% \times 80$$

3. **Solve for x:**

$$x = 0.20 \times 80$$

$$
\begin{array}{r}
80 \\
\times\ 0.20 \\
\hline
0 \\
1600 \\
\hline
16.00
\end{array}
$$

Present the following word problem to your child:

> Marcia is a good softball player. In the last four games, Marcia got 7 hits in 14 at-bats. What percentage of the time does Marcia get a hit?

1. **Write the question asked:**

   7 is what % of 14?

2. **Rewrite the question as an expression:**

   $7 = x\%$ of 14

3. **Solve for x:**

   $$7 = x\% \text{ of } 14$$
   $$7 = x\% \times 14$$
   $$\frac{7}{14} = x\%$$
   $$0.50 = x\%$$
   $$50\% = x$$

Present the following word problem to your child:

> David wants to buy a shirt priced at $50 that has a 25% discount. How much will David pay for the shirt?

1. **Determine what percentage David will pay for the shirt.**

   $100\% - 25\% = 75\%$

2. **Write the question asked:**

   What is 75% of 50?

3. **Rewrite the question as an expression:**

   $x = 75\% \times 50$

4. **Solve for x:**

$$
\begin{array}{r}
50 \\
\times\ 0.75 \\
\hline
250 \\
3500 \\
\hline
37.50
\end{array}
$$

Present the following problem to your child:

Billy's basketball team made 25% of their shots. They made 7 shots. How many shots did Billy's team take?

1. **Write question asked:**

   7 is 25% of what number?

   $7 = 25\% \times x$

2. **Solve for x:**

   $7 / 0.25 = x$

   $28 = x$

3. **Say to your child, "Billy's team took 28 shots."**

Chapter **16**

Making Sense of the Metric System

understand inches and feet. I grew up with them and I can cut wood to the nearest inch and when I want cereal for breakfast, I grab a gallon of milk from the fridge. This system works for me. Given that I don't plan to move to Europe any time soon, why do I need to understand the metric system?

This chapter examines the metric system and addresses why you should care about it. So, turn off the football game, put down your feet, and get ready to understand meters, liters, and grams.

The Need for a Metric System

Like I was saying, I know how to use inches and feet. There are 12 inches to a foot and 3 feet to a yard. Everyone knows that! If I need to measure long distances, I use miles, each of which consists of 5,280 feet.

When I buy milk, I can buy a gallon, and if I need less, I can buy a quart — you remember, there are 4 quarts to a gallon. And, if I want even less, I can buy a pint — there are 2 of those in a quart and 8 of them in a gallon.

Further, when I need to know the weather, I can use temperatures in degrees Fahrenheit, where water freezes at 32 degrees and boils at 212.

What could possibly make more sense? Well, maybe a metric system that is based on decimal math!

Decimal math is base-10 math, with ones, tens, hundreds, and thousands. Decimal math also has tenths, hundredths, and thousands. It's the system the world uses for math.

The metric system is based on decimal math. This means that it specifies distances consistently in terms of meters, liquids in terms of liters, weights in terms of grams, and temperatures in terms of Celsius.

In the metric system, there are no pints, quarts, or gallons. Likewise, there are no inches, feet, yards, or miles. Nothing to memorize, just decimal math.

If you are beginning to think such a system could make sense, you are right!

Covering the Metric Units Your Kid Needs to Know

Your child should know these metric units of measure:

>> Distance is measured in meters

>> Liquids are measured in liters

>> Weight is measured in grams

>> Temperatures are measured in Celsius

To start, you should introduce your child to the metric system.

Say to your child: "You have learned to measure items in terms of inches and feet and to weigh things in terms of pounds. In the United States, we use inches, feet, and pounds as our units of measure. Outside of the United States, the rest of the world measures things using the 'metric' system."

Say to them: "You are going to learn the basics of the metric system. To start, you will learn the metric units of measure."

Present Table 16-1 to your child.

TABLE 16-1

The Metric System Units of Measure

Item to Measure	Metric Units of Measure
Distance	Meters
Liquid	Liters
Weight	Grams
Temperatures	Celsius

Explain to your child: "To measure distances, the metric system uses meters. Likewise, to measure liquids, it uses liters. To measure weight, the metric system uses grams, and to measure temperature, it uses degrees Celsius."

TIP

Buy a metric tape measure or a meter stick (yes, they sell them on Amazon). Let your child measure things, including themselves, in terms of meters.

Say to your child: "The advantage of using the metric system is when you do the math. Rather than having to work in terms of inches, feet, and miles, everything in the metric system is measured in terms of meters."

Spread your arms apart a little wider than 3 feet (there we go again, trying to put things in terms of feet and inches!). Tell your child that a meter is about this big. You might tell them that they are about 1 meter and a half tall.

Explain to your child: "To measure things that are bigger than a meter, you use multiples, such as 2 meters, 5 meters, or 100 meters. To measure things that are smaller than a meter, you use decimal values, such as 0.5 meter, 0.01 meter, and so on."

Present Table 16-2 to your child.

Explain to your child: "To specify metric values, you use these prefixes. For example, 1,000 meters, which is longer than 3 football fields, is a kilometer. The length of your finger is about 10 centimeters."

TABLE 16-2

The Metric System Prefix Values

Prefix	Value
Kilo	1,000
Hecto	100
Deka	10
Deci	0.1 or $\frac{1}{10}$
Centi	0.01 or $\frac{1}{100}$
Milli	0.001 or $\frac{1}{1000}$

TIP

Following are some easy ways to work the study of the metric system into your child's everyday life:

>> If you have a metric tape measure, allow your child to measure items in terms of meters and centimeters. Remind your child that a centimeter is $\frac{1}{100}$ of a meter.

>> If you have a bottle of soda, check to see if it specifies the number of liters on the label. If so, show it to your child, reminding them that the metric system specifies the amount of liquid using liters.

>> The next time you drive with your child in your car, ask them how many meters there are in a kilometer. Then, use your speedometer to show them how fast you are traveling in terms of kilometers. They'll think you are going fast!

>> If your bathroom scale can toggle between pounds and grams, toggle it to grams and let your child weigh themselves and other objects.

>> The next time you watch a track meet (you can find one on ESPN), point out to your child that the races are often based on metric distances, such as the 100-meter dash or the 400-meter dash.

>> If you happen to have a metric toolset, show your child the tools and tell your child that their sizes are expressed in centimeters.

Water Freezes at 0 and Boils at 100 — Introducing Your Kid to Celsius

In the United States, when we talk about temperatures, we use degrees Fahrenheit. The rest of the world, however, uses degrees Celsius. In this section, you introduce your child to Celsius. They also learn how to convert between the two systems — a common algebra problem.

You may be thinking: "What's wrong with using degrees Fahrenheit?" After all, we all know that water freezes at 32 degrees, some people know that water boils at 212 degrees, and when you were a kid, if you could get your temperature from 98.6 to 100 degrees, you may have been able to stay home from school! Why then, should you worry about Celsius?

It turns out that, like the rest of the metric system, Celsius is based on decimal math. Using Celsius, water freezes at 0 degrees — that makes some sense — and water boils at 100 degrees. That means that 0 degrees Celsius is cold (unless, of course, you live in Minnesota where the winter temperatures are often much colder than just freezing — yes, temperatures in Celsius can be below 0). Likewise, using Celsius, 100 degrees is hot, very hot, boiling water hot!

Say to your child: "You have learned that using the metric system, you measure distances using meters, liquids using liters, and weights using grams. In addition, with the metric system, you also measure temperatures using degrees Celsius. The United States, unlike the rest of the world, does not use degrees Celsius — it uses degrees Fahrenheit."

Explain to your child: "Celsius bases temperatures on two key things: the temperature at which water freezes and becomes ice and the temperature at which water boils. With Celsius, water freezes at 0 degrees and water boils at 100 degrees."

Also explain to them: "In contrast, using Fahrenheit, water freezes at 32 degrees and boils at 212 degrees."

Say to your child: "As you start to learn algebra, a common problem you will encounter is converting between degrees Celsius and Fahrenheit. To do so, you use the following expressions."

$$\text{Degrees Celsius} = \frac{5}{9} \times (\text{Degrees Fahrenheit} - 32)$$

$$\text{Degrees Fahrenheit} = \left(\frac{9}{5} \times \text{Degrees Celsius}\right) + 32$$

Consider the following example:

In Arizona, the summers are hot and the temperatures are often over 100 degrees Fahrenheit every day. A visitor from Canada (where they use degrees Celsius) wants to know how hot 100 degrees Fahrenheit is in degrees Celsius.

1. **Present the following expression to your child:**

   $$\text{Degrees Celsius} = \frac{5}{9} \times (\text{Degrees Fahrenheit} - 32)$$

2. **Substitute 100 degrees for "Degrees Fahrenheit" in the expression:**

   $$\text{Degrees Celsius} = \frac{5}{9} \times (100 - 32)$$

3. **Solve the expression:**

   $$\frac{5}{9} \times (100 - 32) =$$
   $$\frac{5}{9} \times 68 = 37.8°$$

4. **Explain to your child that 100 degrees Fahrenheit equals 37.8 degrees Celsius.**

Present the following problem to your child:

In Celsius, water freezes at 0 degrees. Convert 0 degrees Celsius to degrees Fahrenheit to show the temperature at which water freezes in Fahrenheit.

1. **Present the following expression to your child:**

   $$\text{Degrees Fahrenheit} = \frac{9}{5} \times \text{Degrees Celsius} + 32$$

2. **Substitute 0 degrees for "Degrees Celsius" in the expression:**

   $$\text{Degrees Fahrenheit} = \frac{9}{5} \times 0 + 32$$

3. **Solve the expression:**

   $$\frac{9}{5} \times 0 + 32 =$$
   $$0 + 32 = 32°F$$

4. **Explain to your child that 0 degrees Celsius equals 32 degrees Fahrenheit.**

Working through Metric Word Problems

Two advantages of the metric system are easier math and consistency of measurements. In this section, your child solves word problems based on the metric system.

Present the following word problem to your child:

> On Saturday, Jamie and Mary rode their bikes 10 kilometers. On Sunday, they rode 15 kilometers. How far did Jamie and Mary ride?

Your child should get:

$$\begin{array}{r} 10 \\ + 15 \\ \hline 25 \end{array} \text{ kilometers}$$

Present the following word problem to your child:

> Javier and Luisa weighed themselves using a metric scale. Javier weighed 36 kilograms and Luisa weighed 30 kilograms. What is their average weight in kilograms?

Your child should get:

$$\begin{aligned} \text{Average} &= \frac{(36+30)}{2} \\ &= \frac{66}{2} \\ &= 33 \text{ kilograms} \end{aligned}$$

Present the following word problem to your child:

> Tim is performing a science experiment and must boil water. Tim knows that water boils at 100 degrees Celsius, but he only has a thermometer that measures degrees Fahrenheit. Use the following equation to convert 100 degrees Celsius to degrees Fahrenheit:

$$\text{Degrees Fahrenheit} = \frac{9}{5} \times (\text{Degrees Celsius} + 32)$$

Your child should get:

$$\text{Degrees Fahrenheit} = \frac{9}{5} \times (\text{Degrees Celsius} + 32)$$
$$= \frac{9}{5} \times 100 + 32$$
$$= 1.8 \times 100 + 32$$
$$= 180 + 32$$
$$= 212 \text{ degrees}$$

Present the following problem to your child:

Mary ran 1 mile. There are 5,280 feet in a mile. There are 3.3 feet in a meter. How many meters did Mary run?

Your child should get:

$$\text{Numbers of meters} = 5280 \div 3.3$$
$$= 1600$$

Present the following problem to your child:

Billy weighs 90 pounds. One pound is equal to 0.453 kilogram. How much does Billy weigh in kilograms?

Your child should get:

$$\text{Weight in kilograms} = 90 \times 0.453$$
$$= 40.77$$

3

Succeeding at Seventh-Grade Math

Chapter **17**

Revisiting Ratios with Unit Rates

A *unit rate* is a special type of ratio — you remember ratios, you taught them to your child in Chapter 12 — that looks at a single item, a unit. You use unit rates all the time: miles per gallon of gas, the cost of a single item, the amount of time it takes to recharge an electric car, or how much you are being paid per hour.

In this chapter, you teach your child how to calculate unit rates. By understanding unit rates, your child will better appreciate the savings of a $5 footlong sandwich, how much time they spend playing a video game per day, or just how far an electric car can travel on a single charge.

Calculating Unit Rates

While Chapter 12 focuses on calculating ratios to compare two things, this section focuses on calculating unit rates — a special type of ratio.

Say to your child: "When we go grocery shopping, I often want to compare prices. To do so, I calculate something called the unit cost, which I can then compare to other options."

Explain to them: "Sometimes calculating the unit cost is easy. If, for example, two gas stations are on opposite sides of the street and one is selling gas for $5 and the other is selling it for $5.25, you can quickly determine the best price."

Say to them: "Often, however, your best choice will not be so clear. Consider the following scenario: A grocery store has 100 paper plates for sale for $3.00 and 150 plates for $4.00. How do you determine the best buy?"

Say to your child: "To solve this problem, you must know the unit cost — in other words, the cost per plate."

1. Remind your child that $3.00 is the same as 300 cents.

2. To determine the unit cost for 100 plates at $3.00, you use:

$$100\overline{)300}\ \ ^{3}$$

3. To calculate the unit cost for 150 plates at $4.00, you use:

$$\begin{array}{r} 2.\overline{6} \\ 150\overline{)400.0} \\ \underline{300} \\ 1000 \\ \underline{900} \\ 100 \end{array}$$

4. Explain to your child that in this case, 2.6 cents per paper plate is a better buy than 3 cents a plate. Thus, 150 plates for $4.00 is a better buy.

Present the following scenario to your child:

Bill and Ted both rented cars for the weekend. Bill's Camaro went 340 miles on 20 gallons of gas. Ted's Camry went 280 miles on 14 gallons of gas. Which car got better gas mileage?

Say to your child: "In this case, you must determine miles-per-gallon — the unit rate."

1. Explain to your child that you must determine each car's miles per gallon (MPG).

2. To determine the MPG for Bill's Camaro, divide the number of miles he drove, 340, by the number of gallons used:

$$20\overline{)340} \quad \begin{array}{r} 17 \\ \underline{340} \\ 0 \end{array}$$

3. Do the same to determine the MPG for Ted's Camry:

$$14\overline{)280} \quad \begin{array}{r} 20 \\ \underline{280} \\ 0 \end{array}$$

4. Say to your child: "In this case, Ted's Camry, at 20 miles per gallon, got the better mileage."

Explain to your child: "The unit rate is so named because the goal is to determine the rate for one unit. This could be the cost of one paper plate or the number of miles for one gallon of gas."

Ask your child to consider the following example:

Jim's dad paid $100 for 20 gallons of gas. What is the unit rate for the gas — meaning, what is the price for a gallon of gas?

1. Ask your child to specify the ratio of dollars to gas:

$$\frac{100}{20} \text{ dollars} \qquad \frac{100}{20} \quad \text{or} \quad 100{:}20$$

2. Using proportional ratios, as discussed in Chapter 12, you get:

$$\frac{100}{20} = \frac{?}{1}$$

3. Have your child cross-multiply the numbers, first multiplying 100 by 1:

$$100 \times 1 = 100$$

4. Have your child divide the 100 by 20:

$$20\overline{)100} \quad \begin{array}{r} 5 \\ \underline{100} \\ 0 \end{array}$$

5. Say to your child: "In this case, Jim's dad paid $5.00 per gallon of gas."

Present the following problem to your child:

Marsha bought a cheese pizza with 8 slices for $10. How much did Marsha pay per slice?

1. **Ask your child to specify the ratio of dollars to slices:**

$\dfrac{10}{8}$ dollars $\quad\quad$ $\dfrac{10}{8}$ $\quad$ or $\quad$ 10:8
$\phantom{\dfrac{10}{8}}$ slices of pizza

2. **Using proportional ratios, you should get:**

$$\dfrac{10}{8} = \dfrac{?}{1}$$

3. **Have your child cross-multiply the numbers, first multiplying 10 by 1:**

$10 \times 1 = 10$

4. **Have your child divide 10 by 8:**

$$
\begin{array}{r}
1.25 \\
8\overline{)10.00} \\
\underline{8} \\
20 \\
\underline{16} \\
40
\end{array}
$$

5. **Tell your child that Marsha paid $1.25 per slice.**

COMMON UNIT RATES

You use unit rates every day. Here are just a few:

- Miles per gallon
- Credit card interest rates per year
- Cost per pound or item
- Speed limits in miles per hour
- Calories per serving
- Points per game

Working Through Unit-Rate Word Problems

Throughout this chapter, you have helped your child solve unit-rate problems. In this section, they solve word problems that require unit rates.

Present the following word problem to your child:

A pizzeria sells 3 pizzas for $18. What is the cost per pizza?

1. **Have your child create the ratio of dollars per pizza:**

$\dfrac{18}{3}$ dollars pizzas $\dfrac{18}{3}$ or   18:3

2. **Have your child create the proportional ratio:**

$\dfrac{18}{3} = \dfrac{?}{1}$

3. **Have your child cross-multiply the numbers, first multiplying 18 by 1:**

$18 \times 1 = 18$

4. **Have your child divide 18 by 3:**

$$\begin{array}{r} 6 \\ 3{\overline{)18}} \\ \underline{18} \\ 0 \end{array}$$

5. **Tell your child that the pizzeria sells pizzas for $6.00 each.**

Present the following word problem to your child:

The pet store sells a can of dog food for $1.50 and 6 cans for $8.00. Which is the better buy?

1. **Calculate the unit cost for 1 can of dog food for $1.50:**

1 can costs $1.50.

2. **Calculate the unit cost for 6 cans of dog food for $8.00:**

$$
\begin{array}{r}
1.3\overline{3} \\
6)\overline{8.0} \\
\underline{6} \\
20 \\
\underline{18} \\
20 \\
\underline{18} \\
2
\end{array}
$$

1 can costs about $1.33.

3. **Point out to your child that 6 cans of dog food for $8.00 is the better buy.**

Present the following problem to your child:

Sarah and her dad need to put chains on 3 gates to keep in Sarah's horses. Each gate needs a 2-foot chain. Chains cost $0.50 an inch. How much will Sarah and her dad spend on chains?

1. **Calculate the amount of chain they need:**

3 gates × 2 feet = 6 feet

2. **Convert the number of feet to inches:**

6 feet × 12 inches per foot = 72 inches

3. **Determine the cost:**

72 inches × $0.50 per inch = $36

Chapter **18**

Converting Fractions to Decimals

Fractions never seem to go away! At this point, your child should know how to add, subtract, multiply, and divide them! Now, they'll find out how to convert them to a decimal equivalent, which is helpful for working with percentages and in quickly determining the value of a "half-off sale!"

In this chapter, you help your child understand that to convert any fraction into its decimal equivalent, you simply divide the bottom number (the denominator) into the top number (the numerator).

So, get ready to do some division to convert fractions!

Knowing Common Fractions

Before you teach your child how to convert fractions to decimal equivalents, there are several common fractions for which your child should know the corresponding decimal equivalents. In this section, you create flashcards your child can practice with.

Using 3x5 index cards, create the following flashcards, writing the fractions on one side and the equivalent decimal value on the back:

Fraction	Decimal Equivalent	Fraction	Decimal Equivalent	Fraction	Decimal Equivalent
$\frac{1}{2}$	0.5	$\frac{1}{4}$	0.25	$\frac{3}{4}$	0.75
$\frac{1}{3}$	0.33	$\frac{2}{3}$	0.66	$\frac{1}{10}$	0.1
$\frac{2}{10}$	0.2	$\frac{3}{10}$	0.3	$\frac{4}{10}$	0.4
$\frac{5}{10}$	0.5	$\frac{6}{10}$	0.6	$\frac{7}{10}$	0.7
$\frac{8}{10}$	0.8	$\frac{9}{10}$	0.9	$\frac{1}{8}$	0.125
$\frac{3}{8}$	0.375	$\frac{5}{8}$	0.625	$\frac{7}{8}$	0.875
$\frac{1}{5}$	0.2	$\frac{2}{5}$	0.4	$\frac{3}{5}$	0.6
$\frac{4}{5}$	0.8	$\frac{1}{6}$	0.167	$\frac{5}{6}$	0.833

TIP

Practice the fraction flashcards on a regular basis until your child masters them. Then, use the flashcards to present the decimal values and have your child name the equivalent fraction.

FIND ONLINE

This book's companion website at www.dummies.com/go/teachingyourkids newmath6-8fd contains a worksheet that your child can use to practice common fraction and decimal value equivalents. Download and print the worksheet. Help your child with the first few and then ask them to complete the rest.

Converting Any Fraction to Its Decimal Equivalent

In the previous section, your child learned common fractions and their decimal equivalents. In this section, you teach your child how to convert any fraction to its decimal equivalent.

As you may recall, a fraction has a *numerator* (top number) and a *denominator* (bottom number). To convert any fraction to its decimal equivalent, you simply divide the numerator by the denominator. Consider the following:

$$\frac{1}{2} \qquad \frac{1}{4} \qquad \frac{1}{3}$$

By dividing the fractions, you get the following decimal equivalents:

$$\frac{1}{2} \qquad \frac{1}{4} \qquad \frac{1}{3}$$

$$
\begin{array}{r}
0.5 \\
2\overline{)1.0} \\
\underline{10} \\
0
\end{array}
\qquad
\begin{array}{r}
0.25 \\
4\overline{)1.00} \\
\underline{8} \\
20 \\
\underline{20} \\
0
\end{array}
\qquad
\begin{array}{r}
0.3\overline{3} \\
3\overline{)1.00} \\
\underline{9} \\
10 \\
\underline{9}
\end{array}
$$

Say to your child: "You have learned the common fractions and their decimal equivalents. Now, you will learn to convert any fraction to its decimal equivalent."

Explain to your child: "A fraction consists of two parts: a numerator (top number) and a denominator (bottom number). To convert any fraction into its decimal equivalent, you simply divide the fraction's numerator by its denominator."

Consider the following:

$$\frac{3}{4}$$

Say to your child: "To calculate the equivalent decimal value, divide 4 into 3."

$$
\begin{array}{r}
0.75 \\
4\overline{)3.00} \\
\underline{28} \\
20 \\
\underline{20} \\
0
\end{array}
$$

Tell your child: "The decimal equivalent of $\frac{3}{4}$ is 0.75."

Ask your child to convert the following fractions to their decimal equivalent values:

$$\frac{2}{3} \qquad \frac{4}{5} \qquad \frac{1}{8}$$

Your child should get:

$$\frac{2}{3} \qquad\qquad \frac{4}{5} \qquad\qquad \frac{1}{8}$$

```
      0.66̄              0.8               0.125
  3)2.00            5)4.0             8)1.000
    18                 40                  8
    ‾‾                 ‾‾                 ‾‾
     20                  0                 20
     18                                    16
     ‾‾                                    ‾‾
      2                                     40
                                           40
                                          ‾‾
                                            0
```

If they do not, review the division operations with your child.

Ask your child to consider the following fraction:

$$\frac{21}{70}$$

Explain to your child: "To convert the fraction $\frac{21}{70}$ into its decimal equivalent, you simply divide 70 into 21."

```
        0.3
  70)21.0
     210
     ‾‾‾
       0
```

Ask your child to repeat this process for the following fraction:

$$\frac{237}{400}$$

They should get:

```
          0.5925
  400)237.000
      2000
      ‾‾‾‾
      3700
      3600
      ‾‾‾‾
      1000
       800
      ‾‾‾‾
      2000
      2000
      ‾‾‾‾
         0
```

This book's companion website at www.dummies.com/go/teachingyourkids newmath6–8fd contains a worksheet that your child can use to practice converting fractions to their decimal equivalents. Download and print the worksheet. Help your child solve the first few problems and then ask them to complete the rest.

For this worksheet exercise, your child's focus should be on the process of converting a fraction into a decimal, rather than practicing long division. If your child can successfully solve the first few problems without a calculator, you can let them use a calculator to complete the rest.

Word Problems that Convert Fractions to Decimals

In this section, your child gets a chance to solve real-world problems that require them to convert fractions to decimals.

Present the following problem to your child:

> Sarah and Monique are shopping for clothes. Sarah finds a blouse that is half off of $30.00. How much does the blouse cost?

1. **Convert the fraction $\frac{1}{2}$ to a decimal value:**

$$\frac{1}{2} = 0.50$$

2. **Multiply the discount times the original price of $30.00.**

Price $= 0.50 \times \$30$

$$\begin{array}{r} 30 \\ \times\ 0.5 \\ \hline \$15.00 \end{array}$$

Present the following problem to your child:

> Jimmy motorcycle's gas tank holds 5 gallons of gas. Jimmy currently has $\frac{1}{4}$ of a tank. Gas costs $5.25 a gallon. How much will Jimmy spend for gas?

1. **Determine how much gas Jimmy needs:**

   Full tank $= 1 - \dfrac{1}{4} = \dfrac{3}{4}$ of a tank

2. **Convert $\dfrac{3}{4}$ to a decimal.**

   $\dfrac{3}{4} = 0.75$

3. **Determine how many gallons of gas Jimmy needs:**

   Gallons needed $= 0.75 \times 5$

   $$\begin{array}{r} 5 \\ \times\, 0.75 \\ \hline 3.75 \end{array}$$

4. **Multiply the gallons needed times the price of gas:**

   Cost $= 5.50 \times 3.75$

   $$\begin{array}{r} 5.50 \\ \times\, 3.75 \\ \hline \$20.63 \end{array}$$

REAL-WORLD EXAMPLES OF CONVERTING FRACTIONS TO DECIMALS

If you (or you kid) are wondering why it's important to know how to convert fractions to decimals, here are few ways you use this skill in everyday life:

- Understanding how much is half off
- Knowing a half hour is 30 minutes
- Simplifying math: $\dfrac{1}{3} + \dfrac{1}{4} = 0.33 + 0.25 = 0.58$
- Knowing when you should stop for gas
- Setting the right amount of time on the microwave

Chapter **19**

Properties of Operations

I f you are ever on social media, you may have friends who post math expressions to remind everyone how smart they are — something like:

$$3 - 2 \times 6 \div 2$$

Most people will say the answer is either 2 or −3:

$3 - 2 \times 6 \div 2$	or	$3 - 2 \times 6 \div 2$
$= 1 \times 6 \div 3$		$= 3 - 12 \div 2$
$= 6 \div 3$		$= 3 - 6$
$= 2$		$= -3$

If you are in the group that said −3, you are right! If you said 2, this chapter will be a great refresher!

This chapter examines operator precedence that tells you which math operations to perform first. Then, the chapter looks at several key properties of operations your child will use as they encounter more advanced expressions.

Revisiting the Order of Operations — Good Ol' PEMDAS

When you perform arithmetic operations, the order in which you perform the operations matters. Here is the order you should follow:

1. Solve expressions that are grouped within parentheses.
2. Solve exponents.
3. Perform multiplication and division operations from left to right.
4. Perform addition and subtraction operations from left to right.

TIP

You can easily remember the order of operations by using the acronym PEMDAS, which stands for: *Parentheses, Exponents, Multiplication,* and *Division* from left to right, and *Addition* and *Subtraction* from left to right.

In this section, you review the order of operations with your child.

Ask your child to solve the following expression:

$$8 - 5 \times 2 - 4 \div 2 =$$

If they apply the operations in the correct order, they should get -4. If they do not, remind them of the order of operations and then walk them through the following steps:

$$8 - 5 \times 2 - 4 \div 2 =$$
$$8 - 10 - 4 \div 2 =$$
$$8 - 10 - 2 =$$
$$-2 - 2 =$$
$$-4$$

Ask your child to solve the following expression:

$$2 \times (5 - 1) + 8 \div 4 =$$

This time, they should get 10. If they do not, walk them through the following steps:

$$2 \times (5 - 1) + 8 \div 4 =$$
$$2 \times 4 + 8 \div 4 =$$
$$8 + 8 \div 4 =$$
$$8 + 2 =$$
$$10$$

Present the following problem to your child:

$$(2+1)^2 \times 3 =$$

Your child should get:

$$(2+1)^2 \times 3 =$$
$$3^2 \times 3 =$$
$$9 \times 3 =$$
$$27$$

**FIND
ONLINE**

This book's companion website at www.dummies.com/go/teachingyourkids newmath6–8fd contains a worksheet that your child can use to review the order of arithmetic operations. Download and print the worksheet. Help your child solve the first few and then ask them to complete the rest.

Identity Property of Addition

In this section, you teach your child about the identity property of addition, which is a fancy way to say that 0 plus any number, is that number:

5	10	100
+ 0	+ 0	+ 0
5	10	100

Say to your child: "You know that any number plus 0 is equal to that number."

8	25	1000
+ 0	+ 0	+ 0
8	25	1000

Explain to them: "It turns out that mathematicians have a fancy name for this rule, called the *identity property of addition*. The property is so named because when you add 0 to any number, the number keeps its identity (value)."

Identity Property of Multiplication

In this section, you teach your child about the identity property of multiplication — again, a fancy way to say, "Any number times 1 is that number."

8	25	1000
×1	×1	× 1
8	25	1000

Say to your child: "You know that any number times 1 is that number:"

$$
\begin{array}{ccc}
5 & 10 & 100 \\
\underline{\times 1} & \underline{\times 1} & \underline{\times 1} \\
5 & 10 & 100
\end{array}
$$

Explain to them: "Mathematicians call this rule the *identity property of multiplication* because when you multiply any number times 1, the number retains its identity (value)."

Multiplicative Property of Zero

In this section, you teach your child the multiplicative property of zero, which tells you that any number times zero is zero.

$$
\begin{array}{ccc}
8 & 25 & 1000 \\
\underline{\times 0} & \underline{\times 0} & \underline{\times \ 0} \\
0 & 0 & 0
\end{array}
$$

Say to your child: "You know that zero times any number equals zero."

$$
\begin{array}{ccc}
4 & 15 & 1200 \\
\underline{\times 0} & \underline{\times 0} & \underline{\times \ 0} \\
0 & 0 & 0
\end{array}
$$

Explain to them: "Mathematicians call this rule the *multiplicative property of zero*. Note that the name does not use the term 'identity' because when you multiply a number times zero, the result is not equal to the number's original identity (value)."

Commutative Property of Addition

Order is important — but not always! In this section, you teach your child about the commutative property of addition, which says that changing the order in which you add values does not change the sum.

$$
\begin{array}{cc}
1+3=3+1 & 2+4=4+2 \\
4=4 & 6=6
\end{array}
$$

TIP

For this purpose, you can think of the word *commutative* as meaning 'exchangeable.'

Say to your child: "When you add numbers, you normally do so from left to right."

$$3+5+2=10$$

Tell them: "It turns out, however, that you can actually add the numbers in any order to get the same result."

$$3+5+2=10$$
$$2+5+3=10$$
$$5+2+3=10$$

Explain to them: "Mathematicians refer to this rule as the *commutative property of addition*. For simplicity, you can think of it as the exchangeable property of addition."

Associative Property of Addition

You know that when an expression has parentheses, you solve the operations within the parentheses first:

$$(3+5)+1=8+1 \text{ both equal } 9$$

In this section, you teach your child about the associative property of addition, which states that when an expression contains only addition, how you group the operations does not change the result:

$$(4+3)+1=4+(3+1) \text{ both equal } 8$$

Say to your child: "You know that when an expression has operations grouped in parentheses, you solve those operations first."

$$(3+5)\times 2 =$$
$$8\times 2 =$$
$$16$$

Tell them: "When an expression contains only addition operations, it turns out that how you group (or associate) operations does not matter."

$$(5+2)+3=5+(2+3) \text{ both equal } 10$$

Explain to them: "Mathematicians call this rule the *associative property of addition*. For simplicity, you can think of it as the grouping property of addition."

Commutative Property of Multiplication

Earlier in this chapter, you taught your child about the commutative (or exchangeable) property of addition, which states that the order in which you perform arithmetic operations does not change the result. In this section, you teach your child about the commutative property of multiplication, which says the order in which you perform multiplication operations does not change the product:

$3 \times 5 = 5 \times 3$ both equal 15

Say to your child: "You have learned about the commutative (or exchangeable) property of addition, which states that the order in which you perform addition operations does not change the sum."

$5 + 7 = 7 + 5$ both equal 12

Explain to them: "In a similar way, the *commutative property of multiplication* states that the order in which you perform multiplication operations does not change the result."

$5 \times 4 = 5 \times 4$ both equal 20

Knowing Why Subtraction and Division Are Not Commutative

In the previous sections, you saw that addition and multiplication are commutative, which means the order in which you perform such operations does not change the result. In this section, you teach your child that subtraction and division are not commutative.

Say to your child: "You have learned that addition and multiplication are commutative in that the order in which you perform the operations does not change the result."

$5 + 3 = 3 + 5$ both equal 8 $2 \times 4 = 4 \times 2$ both equal 8

Explain to them: "In contrast, subtraction and division are not commutative. This means that you cannot change the order of the operations and get the same result."

Here are two examples:

$7 - 2 \neq 2 - 7$ $9 \div 3 \neq 3 \div 9$

Associative Property of Multiplication

Earlier in this chapter, you taught your child about the associative property of addition, which states that when an expression contains addition operations, how you associate (group) the operations does not change the sum:

$$(1+2)+3 = 1+(2+3) \quad \text{both equal 6}$$

In this section, you introduce your child to the associative property of multiplication, which states that when an expression contains multiplication operations, how you group (or associate) the operations does not change the product:

$$(2 \times 3) \times 4 = 2 \times (3 \times 4) \quad \text{both equal 24}$$

Say to your child: "You have learned that the associative property of addition states that when an expression has addition operations, how you choose to group (associate) the addition operations does not change the sum."

$$(3+2)+5 = 3+(2+5) \quad \text{both equal 10}$$

Explain to them: "In a similar way, the *associative property of multiplication* states that when an expression contains multiplication operations, how you group the operations does not change the product."

$$(3 \times 4) \times 5 = 3 \times (4 \times 5) \quad \text{both equal 60}$$

Distributive Property of Multiplication

As the expressions your child encounters increase in difficulty, they will eventually solve expressions that multiply a number times the result of an addition or subtraction operation that is grouped in parentheses:

$$5 \times (4+3) = \qquad\qquad 3 \times (4-1) =$$

REMEMBER

According to PEMDAS, you should perform the operations within the parentheses first:

$$5 \times (4+3) = \qquad\qquad 3 \times (4-1) =$$
$$5 \times 7 = \qquad\qquad\quad\; 3 \times 3 =$$
$$35 \qquad\qquad\qquad\qquad\; 9$$

In this section, you teach your child about the distributive property of multiplication, which states that the product of a sum (or difference) is equal to the product of each value taken separately:

$5 \times (4+3) = 5 \times 4 + 5 \times 3$ both equal 35

$3 \times (4-1) = 3 \times 4 - 3 \times 1$ both equal 9

Ask your child to consider the following expression:

$4 \times (3-1) =$

Say to them: "You know that you should first solve the operation within the parentheses."

$4 \times (3-1) =$

$4 \times 2 =$

8

Explain to your child: "Mathematicians have found that when you multiply a number times a sum (or difference), you get the same result if you separately multiply the number times the individual numbers and then perform the addition or subtraction."

Consider the following:

$4 \times (3-1) =$

Say to your child: "The *distributive property of multiplication* states that when you multiply a value times the sum of two values, you will get the same result if you multiply the value times each value individually and then add them."

$4 \times (3-1) = 4 \times 3 - 4 - 1 =$

$4 \times 2 = 12 - 4 = 8$

Tell your child that they will use the distributive property of multiplication as they encounter more advanced problems in algebra.

Chapter **20**

The Geometry of Angles

These days, it seems like everyone has an angle — okay, maybe not everyone, but angles exist. In this chapter, your child examines math angles, which are defined as the space created by two intersecting lines or surfaces.

Your child will learn that angles fall into different categories based on their measure, and by knowing the angle's category, they can often perform simple math operations to calculate the measure of an adjoining angle.

So, find your geometry hat and protractor and get ready to measure some angles!

Revisiting Angles

In previous grades, your child was introduced to angles. In this chapter, you revisit what they have previously learned and then build upon that.

Say to your child: "An *angle* is the space that is defined by two intersecting lines or surfaces."

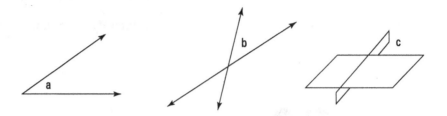

Explain to them: "You normally measure angles in degrees. To measure an angle, you can use a device called a *protractor*, as you can see in Figure 20-1."

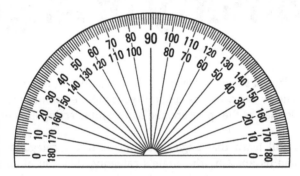

FIGURE 20-1:
Using a
protractor to
measure angles.

Say to your child: "Using the protractor, you can measure the following angles to determine their degrees."

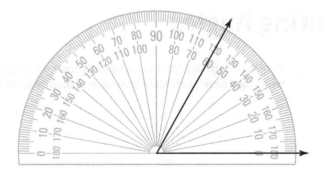

Tell them: "As you will learn, there are mathematical rules about angles that you can use to determine an angle's measure."

Understanding Angle Types

In the previous section, your child learned that you measure angles in degrees. In this section, they learn that you describe angles based on those measurements.

Say to your child: "You describe angles based on their measure — the angle's number of degrees. Based on the angle's measure, you classify angles as a specific type, as you can see here."

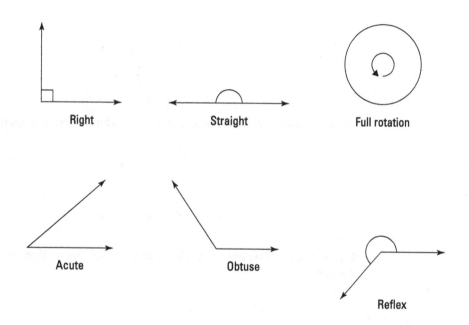

| Right | Straight | Full rotation |
| Acute | Obtuse | Reflex |

Tell your child: "A *right angle* measures 90 degrees. You use the degree (°) symbol to represent degrees. In this case, 90 degrees becomes 90°. A right angle is the angle in each corner of a rectangle or square. You represent a right angle using a small box, as shown here."

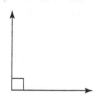

Say to your child: "A *straight angle* forms a straight line that measures 180°."

Explain to them: "An angle can measure from 0° to 360°. A *full-rotation angle* has a measure of 360° and creates a circle."

Tell your child: "An *acute angle* is an angle that measures less than 90°."

Say to them: "Likewise, an *obtuse angle* is an angle that measures greater than 180°."

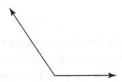

Say to your child: "Finally, a *reflex angle* is an angle that measures greater than 180° and less than 360°."

PREPARING TO WIN AT JEOPARDY!

Some things you should know about angles:

- The angles of a triangle always add up to 180 degrees.

- The angles of a quadrilateral, such as a square or rectangle, always add up to 360 degrees.

- The two angles of the equal sides of an isosceles triangle (two equal sides) are always equal.

- The angles of an equilateral triangle (all equal sides) are always 60 degrees.

- In a triangle, the largest angle is always opposite of the longest side, and the smallest angle is opposite of the shortest side.

Ask your child to identify the following angle types:

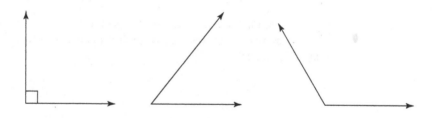

They should get:

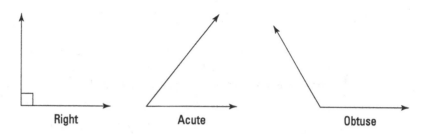

Right   Acute   Obtuse

If they don't, review the angle types with your child.

Understanding Angle Names

To help you refer to specific angles, you can name each angle. Although you can use letters such as a, b, or c, mathematicians often use the Greek symbols alpha, beta, and gamma:

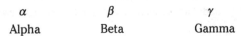

In this section, you teach your child about angle names.

Present the following angles to your child:

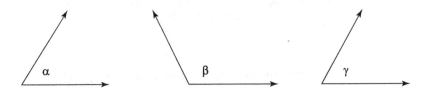

Say to your child: "To help you refer to specific angles, you can label angles using a letter or symbol to represent the angles' names. The following angles use the names *a*, *b*, and *c*."

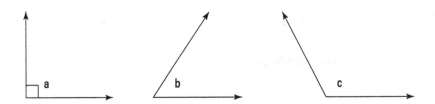

Explain to your child: "Because this is math and many math concepts originated with the Greeks, you will find that mathematicians often use the Greek symbols alpha, beta, and gamma to represent angles."

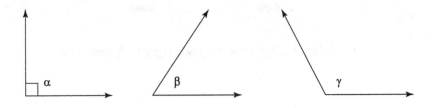

Tell your child: "Whether you use a, b, c, or the Greek symbols, you just need to know that each one represents the angle's name."

Understanding Angle Relationships

Earlier in this chapter, you learned that you can use a protractor to measure angles. In addition, you can often calculate one angle's value if you know the measure of an angle next to it. In this section, you teach your child how to do just that.

Ask your child to consider the following angle:

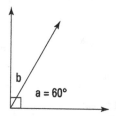

Say to your child: "In this case, you have two angles that collectively form a 90° right angle. You know that the measure of angle a is 60° but you do not know the measure of angle b."

Explain to them: "In math, two angles that add to 90° are called *congruent angles*. Knowing that your two angles equal 90°, you can calculate angle b's measure as follows."

$$b = 90° - a$$
$$= 90° - 60°$$
$$= 30°$$

Ask your child to solve for the following unknown angles:

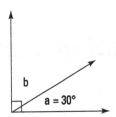 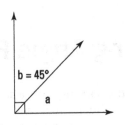 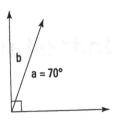

They should get:

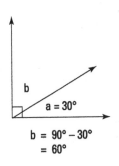 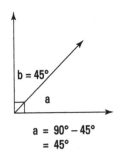 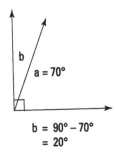

$b = 90° - 30°$ $a = 90° - 45°$ $b = 90° - 70°$
    $= 60°$ $= 45°$ $= 20°$

If they do not, help your child subtract the known angle's measure from 90°.

Present the following angle to your child:

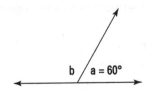

Say to your child: "In this case, you know that angle a is equal to 60° but you don't know the measure for angle b."

Explain to them: "In math, angles that add up to 180° are called *supplemental angles*. Knowing that angles *a* and *b* form a straight line (a 180° supplemental angle), you can solve for angle *b* as follows."

$$b = 180° - a$$
$$= 180° - 60°$$
$$= 120°$$

Ask your child to solve for the following unknown angles:

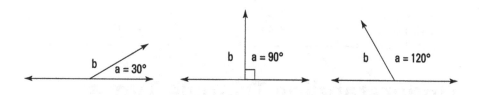

They should get:

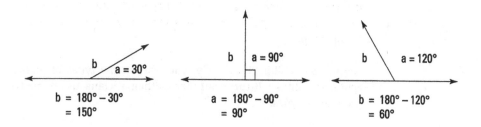

$$b = 180° - 30°$$
$$= 150°$$

$$a = 180° - 90°$$
$$= 90°$$

$$b = 180° - 120°$$
$$= 60°$$

If they do not, help your child subtract the known angle from 180.

FIND ONLINE

This book's companion website at www.dummies.com/go/teachingyourkids newmath6-8fd contains a worksheet (the first part of which is shown in Figure 20-2) that your child can use to solve for unknown angles. Download and print the worksheet. Help your child solve the first few and then ask them to complete the rest.

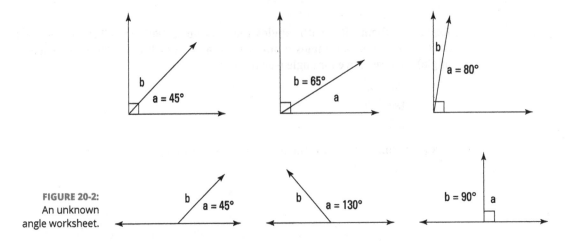

FIGURE 20-2:
An unknown
angle worksheet.

Understanding Triangle Types

In the previous section, your child learned the names of different angle types. As it turns out, *triangles*, three-sided shapes with three angles, have similar names based on their angle types. In this section, your child learns to recognize these triangle types.

Say to your child: "Previously, you learned the types of angles, such as right, acute, and obtuse. It turns out that mathematicians also name triangles based on their angle types."

Tell them: "The following illustrates the triangle types."

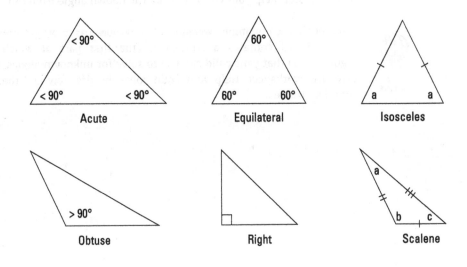

Say to your child: "In an *acute* triangle, all three angles are less than 90°."

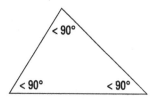

Tell them: "In an *equilateral triangle*, all of the sizes have the same length and each of the angles is 60°."

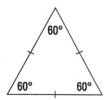

Say to your child: "An *isosceles triangle* has two equal sides and two equal angles."

Tell them: "In an *obtuse triangle*, one angle is greater than 90°."

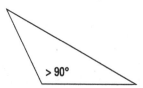

Say to them: "A *right triangle* has a 90° angle, which you represent by placing a small box in the corner."

Tell your child: "Finally, in a *scalar triangle*, all three sides and angles are different."

Ask your child to name the following angle types:

 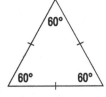

They should get:

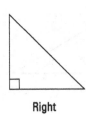 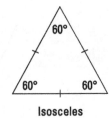 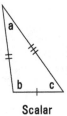

Right   Isosceles   Scalar

If they do not, review each of the triangle types and their angles with your child.

This book's companion website at www.dummies.com/go/teachingyourkids newmath6-8fd contains a worksheet (the first part of which is shown in Figure 20-3) that your child can use to identify triangle types. Download and print the worksheet. Help your child identify the first few and then ask them to complete the rest.

FIGURE 20-3:
A worksheet to identify triangle types.

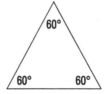
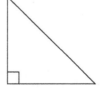

Chapter **21**

Statistics Aren't Perfect

S tatistics is the branch of math that analyzes data. The definition alone can be intimidating and probably makes statistics sound boring! The good news is that your child already knows how to find many common statistics: average value, minimum and maximum values, as well as the median and modal values. It's okay if you don't remember the last two — the *median* is the middle value in a sorted list of data, and the *modal value* is the value that appears most often.

In this chapter, your child finds out how to calculate the *variance*, which is a measure that gives you insights about your data — specifically, how close (or far) the data points are from the average value.

With knowledge of variance, you can impress your friends the next time they tell you about a poll they heard on the evening news. You can tell them that you suspect the poll has a large variance due to outliers, which have skewed the average value!

Why Statistics Aren't Perfect

In the computing world, there's a common saying about data:

Garbage In, Garbage Out

If bad data goes into a computer program, bad results will come out. In a similar way, statistics are only as good as the data that goes into them.

Variance is a measure that provides insight into the quality of data. In this section, you teach your child the whys and hows of variance.

Say to your child: "You have learned to calculate the average value of a set of data."

```
 50
 10            30
 30         5)150
 20           15
 40           00
150
```

Explain to them: "When you calculate the average, your data may all be close to that average, or some data may be far from the average."

Ask your child to consider the following data plots:

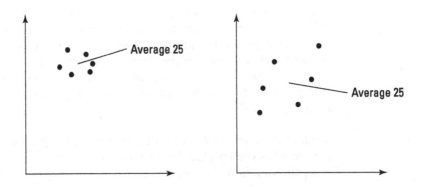

Say to your child: "As you can see, the data in both charts have an average value of 25. In the first chart, all the data is very close to that average value. In the second chart, some of the data is farther away from the average, and in the third chart, the data is quite far from the average."

Explain to them: "Statisticians use the term *variance* to describe how close the data is to the average. In the first chart (with the data close to the average), the variance would be small. In the second chart, the variance would be larger."

Calculating the Variance

In the previous section, your child learned what variance is and also was able to visualize data for which the variance would be small and large. In this section, they learn how to calculate a value for variance.

Say to your child: "You have learned that statisticians use variance as a measure of how close the data is to the average."

Ask your child to consider the following data sets:

Data Set One	Data Set Two
25	50
24	0
23	40
27	10
26	25

1. Say to your child: "To calculate the variance, you must first calculate the average value for each data set."

$$
\begin{array}{cc}
25 & 25 \\
24 & 5\overline{)125} \\
23 & 10 \\
27 & 25 \\
\underline{26} & \underline{25} \\
125 & 0
\end{array}
\qquad
\begin{array}{cc}
50 & 25 \\
0 & 5\overline{)125} \\
40 & 10 \\
10 & 25 \\
\underline{25} & \underline{25} \\
125 & 0
\end{array}
$$

2. Tell your child: "As you can see, the average of each data set is 25. However, the data is very different between the two sets."

3. Explain to them: "To calculate the variance for each data set, you first subtract each data value from the average, then squaring each result."

$$
\begin{aligned}
(25-25)^2 &= (0)^2 = 0 \\
(24-25)^2 &= (-1)^2 = 1 \\
(23-25)^2 &= (-2)^2 = 4 \\
(27-25)^2 &= (2)^2 = 4 \\
(26-25)^2 &= (1)^2 = 1
\end{aligned}
\qquad
\begin{aligned}
(50-25)^2 &= (25)^2 = 625 \\
(0-25)^2 &= (-25)^2 = 625 \\
(40-25)^2 &= (15)^2 = 225 \\
(40-25)^2 &= (15)^2 = 225 \\
(25-25)^2 &= (0)^2 = 0
\end{aligned}
$$

4. Say to your child: "Finally, you calculate the average of each of those squared differences."

0	625
1	625
2	225
2	225
1	0
$\overline{5}$ $5 \div 5 = 1$	$\overline{1700}$ $1700 \div 5 = 340$

Variance $= 1$

5. Explain to them: "As you can see, the variance of the first data set is small, 1, and the variance of the second data set is much larger, 320. If you examine the values in the first data set, you find that each value is close to the average. However, for the second data set, the values are farther away from the average and so there is a large variance."

Ask your child to calculate the variance for the following data sets:

Data Set One	Data Set Two
13	1
17	2
15	2
14	28
16	15

1. Ask your child to compute each data set's average value:

$$
\begin{array}{llll}
13 & 15 & 1 & 17 \\
17 & 5\overline{)75} & 29 & 5\overline{)85} \\
15 & \underline{5} & 2 & \underline{5} \\
14 & 25 & 28 & 35 \\
\underline{16} & \underline{25} & \underline{15} & \underline{35} \\
75 & 0 & 85 & 0
\end{array}
$$

2. Ask your child to subtract each value from the average, squaring the result:

$$
\begin{aligned}
(13-15)^2 &= (2)^2 = 4 \\
(17-15)^2 &= (2)^2 = 4 \\
(15-15)^2 &= (0)^2 = 0 \\
(14-15)^2 &= (-1)^2 = 1 \\
(16-15)^2 &= (1)^2 = 1
\end{aligned}
\qquad
\begin{aligned}
(1-17)^2 &= (-16)^2 = 256 \\
(29-17)^2 &= (12)^2 = 144 \\
(2-17)^2 &= (-15)^2 = 225 \\
(28-17)^2 &= (11)^2 = 121 \\
(15-28)^2 &= (-13)^2 = 169
\end{aligned}
$$

This chapter focuses on calculating the variance for a data set. It turns out that statisticians also use a measure called the *standard deviation* to determine how close data values are to the average value. It turns out that the standard deviation is the square root of the variance. For now, your child simply needs to know how to calculate the variance.

3. **Have your child calculate the average value of the squared differences:**

2		256	
4		144	
0		225	
1		121	
1		169	

$8 \quad 8 \div 5 = 1.6$ $915 \quad 915 \div 5 = 183$

Variance $= 1.6$ Variance $= 183$

Working through Word Problems

In this section, your child solves word problems that require them to calculate the variance of a data set.

Present the following word problem to your child:

The debate team members sold cookies for a week to raise money for an upcoming trip. The following data set shows the team's sales:

Name	Sales
Trevor	15
Mary	25
Javier	5
Alexis	0
Shauna	35

Calculate the variance for the team's cookie sales.

1. Have your child calculate the average value. They should get:

$$
\begin{array}{r}
15 \\
25 \\
5 \\
0 \\
35 \\
\hline
80
\end{array}
\qquad
\begin{array}{r}
16 \\
5\overline{)80} \\
5 \\
\hline
30 \\
30 \\
\hline
0
\end{array}
$$

2. Ask your child to subtract each data value from the average, squaring the results. They should get:

$$(15-16)^2 = (-1)^2 = 1$$
$$(25-16)^2 = (9)^2 = 81$$
$$(5-16)^2 = (-11)^2 = 121$$
$$(0-16)^2 = (-16)^2 = 256$$
$$(35-16)^2 = (19)^2 = 361$$

3. Have your child calculate the average value of the squared differences. They should get:

$$
\begin{array}{r}
1 \\
81 \\
121 \\
256 \\
361 \\
\hline
820
\end{array}
$$

$$820 \div 5 = 164$$
$$\text{Variance} = 164$$

Present the following word problem to your child:

A statistician is interested in the following two data sets. Specifically, they want to compare the variance for each set:

Data Set One	Data Set Two
20	0
80	100
30	2
70	98
50	50

What is the variance of each data set?

1. **Have your child calculate the average value for each data set. They should get:**

$$
\begin{array}{l}
20 \\
80 \\
30 \\
70 \\
\underline{50} \\
250
\end{array}
\qquad
\begin{array}{r}
50 \\
5\overline{)250} \\
\underline{25} \\
00
\end{array}
\qquad
\begin{array}{l}
0 \\
100 \\
2 \\
98 \\
\underline{50} \\
250
\end{array}
\qquad
\begin{array}{r}
50 \\
5\overline{)250} \\
\underline{25} \\
00
\end{array}
$$

2. **For both data sets, ask your child to subtract each data value from the average, applying the absolute value. They should get:**

$$(20-50)^2 = (-30)^2 = 900 \qquad\qquad (0-50)^2 = (-50)^2 = 2500$$
$$(80-50)^2 = (30)^2 = 900 \qquad\qquad (100-50)^2 = (50)^2 = 2500$$
$$(30-50)^2 = (-20)^2 = 400 \qquad\qquad (2-50)^2 = (-48)^2 = 2304$$
$$(70-50)^2 = (20)^2 = 400 \qquad\qquad (98-50)^2 = (48)^2 = 2304$$
$$(50-50)^2 = (0)^2 = 0 \qquad\qquad (50-50)^2 = (0)^2 = 0$$

3. **For each data set, ask your child to calculate the average value of the squared differences. They should get:**

$$
\begin{array}{l}
900 \\
900 \\
400 \\
400 \\
\underline{0} \\
2600 \\
2600 \div 5 = 520 \\
\text{Variance} = 520
\end{array}
\qquad\qquad
\begin{array}{l}
2500 \\
2500 \\
2304 \\
2304 \\
\underline{0} \\
9608 \\
9608 \div 5 = 1921.6 \\
\text{Variance} = 1921.6
\end{array}
$$

IN THIS CHAPTER

» **Understanding what a probability means**

» **Calculating simple probabilities**

» **Realizing that with probabilities, previous results don't matter**

» **Solving combined probabilities for two unrelated events**

» **Calculating probabilities for dependent events**

Chapter **22**

Introducing Probabilities

I f you could only win the lottery — life would be different. To start, you might hand this book off to your newly hired math tutor!

Unfortunately, as you will discover in this chapter, the probability of you winning the lottery is less than your probability of getting struck by lightning. This chapter introduces your child to probabilities, what they mean, and how to calculate them.

So, put down your lottery ticket and let's get started. By the end of this chapter, the probability that you will better understand the math behind probabilities is high!

Understanding Probability

Before you examine how to calculate the probability of an event, it is important that your child understand what probability means. In this section, you teach them just that.

Say to your child: "A *probability* is a value from 0 to 1 that is the likelihood that something will happen. The probability may correspond to you rolling a 7 with the next toss of two dice, being hit by lightning, or winning the lottery."

Explain to your child: "When a probability has a value close to 0, the corresponding event has very, very little chance of happening. In contrast, when the probability value is close to 1, the event is extremely likely to occur. And, when the probability is 0.5, you have a 50/50 chance of the event happening, such as when you flip a coin; half the time you will get heads, and half the time you will get tails."

Present the following illustration to your child:

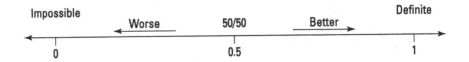

Calculating Simple Probabilities

Once your child understands what probability means, they're ready to learn how to calculate simple probabilities.

Say to your child: "You have learned that a probability is a value from 0 to 1 that indicates the likelihood that an event will occur. Now you will learn how to calculate probability values."

Present a coin to your child, showing that one side is heads and the other is tails.

1. Explain to your child: "To calculate the probability of an event occurring, you divide the total chances of the event happening by the total number of possible outcomes."

2. Say to them: "Assume you want to know the probability of flipping the coin and that it is heads. The coin has only 1 head, so it will become the top number of your fraction. The coin has two possible outcomes (heads or tails), and these become the bottom number."

   Probability of heads $= \dfrac{1}{2}$

3. Explain to them: "The fraction $\dfrac{1}{2}$ equals 0.5, so the probability of flipping a heads is fifty/fifty. Half the times should be heads, and half should be tails."

4. Say to your child: "Consider that a die has the numbers 1 through 6, with one number per side. Assume you want to know the probability of rolling a 1."

5. Explain to them: "The die has only one side with the value 1, which becomes the top number of your fraction. The die has 6 sides (possible outcomes), which means 6 becomes the fraction's bottom number."

Probability of rolling a 1 = $\frac{1}{6}$

6. Say to your child: "To calculate the corresponding probability number, you divide 1 by 6."

$$
\begin{array}{r}
0.16\overline{6} \\
6\overline{)1.000} \\
\underline{6} \\
40 \\
\underline{36} \\
40 \\
\underline{36} \\
4
\end{array}
$$

7. Tell them: "In this case, your probability of rolling the die and having it be 1 is 0.166, or unlikely."

Say to your child: "Assume you want to know the probability of picking a yellow marble from a bag with four blue marbles and one yellow marble — meaning five different colored marbles."

1. Explain to your child: "Since you have only one yellow marble, 1 becomes your fraction's top number. Because you have five different marbles, 5 becomes the fraction's bottom number."

Probability of a yellow marble = $\frac{1}{5}$

2. Tell them: "To determine the probability value, you divide 1 by 5."

$$
\begin{array}{r}
0.2 \\
5\overline{)1.0} \\
\underline{10} \\
0
\end{array}
$$

3. Explain to them: "In this case, your probability of picking a yellow marble from the bag is 0.20, which is unlikely."

4. Say to your child: "Assume, instead, that you want to determine the probability of picking a blue marble from the bag. In this case, you have four blue marbles, which means 4 will be your top number. The total number of marbles is still 5, the bottom number."

Probability of picking a blue marble = $\frac{4}{5}$

5. Say to them: "To calculate the probability, divide 4 by 5."

$$\begin{array}{r} 0.8 \\ 5\overline{)4.0} \\ \underline{40} \\ 0 \end{array}$$

6. Tell your child: "The probability of picking a blue marble from the bag is 0.8, or likely."

Ask your child to consider the following spin wheel:

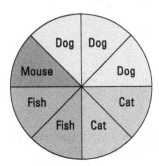

Ask your child: "How many total possibilities does the wheel have?"

They should say 8. If not, count the possibilities with your child.

Ask your child: "What is the probability of spinning the wheel and having it stop on a dog?"

They should get:

$$\frac{\text{Number of dogs}}{\text{Total possibilities}} \quad \frac{3}{8}$$

$$\begin{array}{r} 0.375 \\ 8\overline{)3.000} \\ \underline{24} \\ 60 \\ \underline{56} \\ 40 \\ \underline{40} \\ 0 \end{array}$$

If they don't, count the number of dogs on the spin wheel and help your child per-form the math.

TIP

Ask your child: "What is the probability of spinning a fish?"

They should get:

$$\frac{\text{Number of fish}}{\text{Total possibilities}} \quad \frac{2}{8}$$

$$\begin{array}{r} 0.25 \\ 8\overline{)2.00} \\ \underline{16} \\ 40 \\ \underline{40} \\ 0 \end{array}$$

TIP

Again, if they don't, have them count the number of fish on the wheel and help them with the math.

Finally, ask your child: "What is the probability of spinning a mouse?"

They should get:

$$\frac{\text{Number of mice}}{\text{Total possibilities}} \quad \frac{1}{8}$$

$$\begin{array}{r} 0.125 \\ 8\overline{)1.000} \\ \underline{8} \\ 20 \\ \underline{16} \\ 40 \\ \underline{40} \\ 0 \end{array}$$

Understanding the Impact of Previous Events

In the previous section, your child learned how to calculate simple probabilities. In this section, they learn that such probabilities are not changed by previous results.

Say to your child: "You have learned that the probability of flipping a coin and having it be heads is 50/50. Assume that I flip three heads in a row. Does that change the probability of my next flip?"

The answer is no. The math does not change. You still have 1 head and two possible outcomes:

$$\text{Probability of heads} = \frac{1}{2} \text{ or } 0.5$$

Explain to your child, "Even though I have flipped three heads in a row, the probability of flipping a heads on the next toss is still 0.5. The previous results do not change the probability."

Understanding Combined Probabilities

In the previous section, your child learned how to calculate simple probabilities for single events. In this section, you teach them how to calculate probabilities based on two unrelated events occurring.

Say to your child: "Assume you have a bag of five marbles: blue, green, red, yellow, and orange."

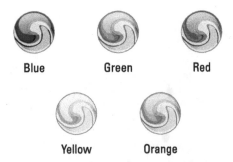

Blue   Green   Red

Yellow   Orange

Tell them: "You first pick a marble that will become the color you use for the following spinner."

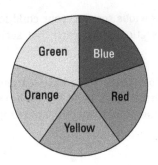

Explain to your child: "You want to determine the probability of picking a blue marble and then spinning the spinner to blue — a combined probability."

1. Say to your child: "First, you must calculate the probability of picking a blue marble. Given that you have 1 blue marble and 5 possibilities, your probability of picking the blue marble becomes the following."

$$\frac{\text{Number of blue marbles}}{\text{Total marbles}} \qquad \frac{1}{5}$$

$$5\overline{)1.0} \quad \begin{matrix} 0.2 \\ 1.0 \\ \underline{10} \\ 0 \end{matrix}$$

2. Say to them: "Next, you must determine the probability of spinning a blue color. Again, you have 1 blue color and 5 possibilities."

$$\frac{\text{Number of blue colors}}{\text{Total colors}} \qquad \frac{1}{5}$$

$$5\overline{)1.0} \quad \begin{matrix} 0.2 \\ 1.0 \\ \underline{10} \\ 0 \end{matrix}$$

3. Explain to your child: "To determine the combined probability of first picking a blue marble and then spinning the color blue, you multiply your two probabilities."

Probability of blue marble $= (0.2) \times (0.2) = 0.04$
and blue spin

4. Say to your child: "The probability of picking the blue marble and then spinning blue is 0.04, which is very, very unlikely."

Understanding Dependent Probabilities

You may be thinking, "Lotto! What's the probability that my five numbers will win the lotto?" In this section, you and your child learn just that.

Explain to your child: "In the previous section, you learned how to calculate the probability of two unrelated events occurring. Now, you will learn how to calculate the probability of two or more dependent events."

Say to your child: "Assume that your bag of marbles has a red, green, blue, yellow, and orange marble — five marbles in total."

Say to them: "You want to know the probability of picking a blue marble and then picking a red marble."

1. Say to your child: "To start, you must first calculate the probability of picking a blue marble from the bag."

   Probability of first picking a blue marble $= \dfrac{1}{5}$

   $$5 \overline{)1.0} \quad 0.2$$
   $$\underline{10}$$
   $$0$$

2. Explain to them: "Next, you need to calculate the probability of picking the red marble second. In this case, you only have 1 red marble (the top number), but because you have removed the blue marble, your total number of possibilities is now 4 (the bottom number)."

   Probability of picking a red marble from remaining four $= \dfrac{1}{4}$

   $$4 \overline{)1.00} \quad 0.25$$
   $$\underline{8}$$
   $$20$$
   $$\underline{20}$$
   $$0$$

3. Say to your child: "To determine the combined probability of picking the blue marble and then picking the red marble, you must multiply the two probabilities."

   Probability of picking blue marble followed by red marble $= (0.2) \times (0.25) = 0.05$

4. Tell them: "In this case, the probability of first picking a blue marble and then picking a red marble from the rest is 0.05, or very, very low."

Say to your child: "Every week, millions of people play the lottery with hopes of winning a big jackpot. To play, the player picks 5 numbers from 1 to 40. Later, the lottery company picks 5 balls from a total of 40 balls numbered 1 through 40."

Tell them: "If the lottery company picks a player's 5 balls, the player can win millions of dollars! Let's calculate the probability of winning the lottery!"

1. Say to your child: "To start, you calculate the probability of picking one of the player's balls on the first pick. In this case, the player has 5 balls (the top number), and the total number of balls is 40."

$$\text{Probability of picking one of five balls} = \frac{5}{40}$$

$$
\begin{array}{r}
0.125 \\
40\overline{)5.000} \\
\underline{40} \\
100 \\
\underline{80} \\
200 \\
\underline{200} \\
0
\end{array}
$$

2. Explain to your child: "Next, you need to calculate the probability of picking one of the player's remaining balls on the next pick. In this case, the player has 4 balls remaining and your total number of possibilities is now 39."

$$\frac{4}{39} = 0.102$$

3. Say to them: "You must repeat this process for the next 3 balls."

$$\frac{3}{38} = 0.079$$

$$\frac{2}{37} = 0.054$$

$$\frac{1}{36} = 0.028$$

4. Tell your child: "Finally, to determine the combined probability of picking all 5 balls, you must multiply each of the probabilities."

$$0.125 \times 0.102 \times 0.079 \times 0.054 \times 0.028 = 0.0000015$$

5. Explain to them: "As you can see, the combined probability of picking all 5 balls is 0.0000015, which is very, very, very unlikely."

Working through Probability Word Problems

You may have been wondering, "What's the probability that this chapter would have word problems?" Your child may be able to help you calculate that.

$$\frac{\text{Chapters with word problems}}{\text{Total chapters}}$$

Present the following word problem to your child:

> Gloria's teacher writes the 26 letters of the alphabet on the chalkboard. They then ask a student to erase any letter. What is the probability that the student will erase a vowel (5 possibilities)?

Your child should get the following:

$$\frac{5}{26} = 0.192$$

Present the following word problem to your child:

> At Javier's birthday party, they had the following spin wheel to determine what they would eat:

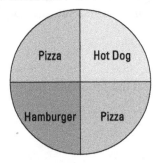

> What is the probability that they would eat pizza?

Your child should get the following:

$$\frac{\text{Number of pizza choices}}{\text{Total possibilities}} = \frac{2}{4} = 0.50$$

Present the following problem to your child:

> Mary got a new camera and wants to photograph geese on the lake. On average, people have seen 8 different geese throughout the day. How long should Mary plan to spend at the lake to get a picture?

Your child should get the following:

DID YOU KNOW?

Your odds of:

- Getting struck by lightning: less than one in a million

- Winning a lottery: less than one in 300 million

- Becoming an astronaut: one in a million

- Being bitten by shark: one in 3.7 million

- Being in a car accident: 8 out of 10

- Guessing Super Bowl coin toss: 1 out of 2

- Becoming a millionaire: about 1 out of 570,000 (worldwide)

- Becoming famous: 1 in 10,000

- Being happy: 6 out of 10

$$\frac{\text{Number of different geese}}{\text{Total hours in a day}} = \frac{8}{24} = 0.33$$

Thus, the probability of a goose being at the lake in any given hour is 0.33. To ensure that she will get a photo, Mary should plan on spending 3 hours at the lake:

$0.33 \times 3 = 1$ (A probability of 1 equals a 100 percent chance of her seeing a goose.)

4

Graduating to Eighth-Grade Math

Chapter **23**

Solving Two-Variable Expressions

You have learned that algebra is a branch of mathematics that uses letters and symbols to represent unknowns in expressions. Your child performs algebra when they solve expressions for the value of *x*:

$$x + 4 = 6$$

In this chapter, your child steps up their algebra skills by learning to solve expressions that have two unknowns: *x* and *y*. To solve for two unknowns, you must start with two expressions — that's a rule, "two unknowns, two expressions." Here are some examples:

$$x + y = 5$$

$$x - y = 1$$

If the examples above "look Greek to you," you can't blame the Greeks—they invented geometry. Algebra can be attributed to ancient Egyptians. Also, relax. You find out how to solve these equations shortly.

So, get ready to have twice the fun with algebra!

Moving from One-Variable to Two Variable Expressions

In Chapter 15, you taught your child to solve one-variable expressions, such as:

$$x = 5 + 1 \qquad 2x = 4 \qquad 3x - 1 = 17 \qquad 3x^2 = 27$$

In this chapter, your child learns to solve problems expressed in terms of two variables, x and y. This means they are looking for two unknowns:

$$x + y = 7$$

$$x - 5 = y$$

REMEMBER

To solve for two unknowns, you must have two equations — that's a rule.

To solve the previous problem, your child can first solve for x. To do so, they can pick one of the expressions. In this case, I'll use:

$$x - 5 = y$$

Your goal is to get x alone on one side of the equal sign. To do that, you can add 5 to both sides of the equal sign:

$$x - 5 = y$$
$$x - 5 + 5 = y + 5$$
$$x = y + 5$$

Next, you can replace the x with $y + 5$ (which is the value of x) in the second expression. Then, you can rewrite the expression from $y + 5 + y = 7$ to $2y + 5 = 7$:

$$x + y = 7$$
$$y + 5 + y = 7$$
$$2y + 5 = 7$$

To get y on its own on one side of the equal sign, you can subtract 5 from both sides and then divide both sides by 2 to solve for y:

$$2y + 5 - 5 = 7 - 5$$
$$2y = 2$$
$$y = 1$$

Now that you know the value of y, which is 1, you can substitute that value into one of the expressions to solve for x:

$$x + y = 7$$
$$x + 1 = 7$$

To get x on its own, you subtract 1 from both sides:

$$x + 1 - 1 = 7 - 1$$
$$x = 6$$

Say to your child: "You have previously learned how to solve expressions for an unknown value of x."

$$x = 3 + 2 \qquad 2x = 6 \qquad 3x - 5 = 7$$

Ask your child to solve these expressions. They should get:

$$x = 3 + 2 \qquad 2x = 6 \qquad 3x - 5 = 7$$
$$x = 5 \qquad \frac{2x}{2} = \frac{6}{2} \qquad 3x - 5 + 5 = 7 + 5$$
$$x = 3 \qquad \frac{3x}{3} = \frac{12}{3}$$
$$x = 4$$

FIND ONLINE

This book's companion website at www.dummies.com/go/teachingyourkids newmath6-8fd contains a worksheet (the first few rows of which are shown in Figure 23-1) that your child can use to solve for x. Download and print the worksheet. Help your child solve the first few and then ask them to complete the rest.

$$x = 5 + 3 \qquad\qquad x = 7 - 2 \qquad\qquad x = 5 \times 5$$

FIGURE 23-1:
A worksheet for using two expressions to solve for x.

$$x - 3 = 7 \qquad\qquad x + 2 = 5 \qquad\qquad x - 3 = 9$$

TIP

Do not move on to two-variable expressions until your child has successfully mastered solving one-variable expressions for x.

Moving Up to Two-Variable Expressions

Often, in algebra, expressions have two unknowns: x and y. To solve for two unknowns, you must have two expressions:

$$x + y = 7$$
$$x - y = 1$$

In this section, you teach your child how to solve for both unknowns.

WHY YOU USE X AND Y

You have seen that you often use *x* and *y* to represent unknowns in expressions, such as $x + y = 5$. You may ask, why *x* and *y*? Why not *a* and *b*?

Credit for why you use *x* and *y* probably goes to the French mathematician, Rene Descartes, who selected *x*, *y*, and *z* for use in a paper written in 1637 because they appear at the end of the alphabet. Beyond that, there's nothing special about the letters. You can represent your unknowns with *a* and *b* — or *c* and *d*, for that matter. However, should you do so, people will likely ask why.

Say to your child: "Now you will learn to solve expressions that have two unknowns, *x* and *y*."

Consider the following expressions:

$$y + x = 10$$

$$x - y = 6$$

Explain to your child: "In this case, you have two variables, *x* and *y*, which are both unknown. To solve for two unknown variables, you must have two expressions. That's a rule!"

Solve for *x* and *y*:

1. **To start, use one of the expressions to solve for *x*:**

   $$x - y = 6$$

2. **Tell your child: "Your goal is to get *x* by itself on one side of the equal sign—called isolating *x*. To do so, you can add *y* to both sides."**

   $$x - y + y = 6 + y$$
   $$x = 6 + y$$

3. **Say to your child: "Now that you know the value of *x*, you can substitute its value (6 + *y*) into one of the expressions."**

   $$y + x = 10$$
   $$y + 6 + y = 10$$
   $$2y + 6 = 10$$

4. **Say to them: "Now you can solve for *y*."**

   $$2y + 6 = 10$$

5. Tell them: "To do so, you first subtract 6 from both sides and then divide by 2 to solve for *y*."

$$2y + 6 - 6 = 10 - 6$$
$$\frac{2y}{2} = \frac{4}{2}$$
$$y = 2$$

6. Say to your child: "Now that you know the value of *y*, which is 2, you can substitute it into one of the expressions to solve for *x*."

$$y + x = 10$$
$$2 + x = 10$$

7. Tell them: "To get *x* alone on one side of the equal sign (to isolate *x*), you subtract 2 from both sides."

$$-2 + 2 + x = 10 - 2$$
$$x = 8$$

8. Say to your child: "The value of *x* is 8 and the value of *y* is 2. If you plug both values into the expressions, you should get the following results."

$$
\begin{array}{ll}
y + x = 10 & x - y = 6 \\
2 + 8 = 10 & 8 - 2 = 6 \\
10 = 10 & 6 = 6
\end{array}
$$

Explain to your child: "In the previous problem, you solved for *x* and *y* by adding or subtracting values from both sides of the equation — meaning, whatever you added or subtracted from one side, you did the same on the opposite side of the equation. As you solve such problems, you may see the addition and subtraction performed beneath the equations."

Consider again, the previous problem.

$$y + x = 10$$

$$x - y = 6$$

1. To start, use one of the expressions to solve for *x*:

$$x - y = 6$$

2. Tell your child: "Your goal is to get *x* by itself on one side of the equal sign — this is called isolating *x*. To do so, you can add *y* to both sides."

$$
\begin{array}{l}
x - y = 6 \\
\underline{+y \quad\quad +y} \\
\quad x = 6 + y
\end{array}
$$

3. Say to your child: "Now that you know the value of *x*, you can substitute its value $(6 + y)$ into one of the expressions."

$$y + x = 10$$
$$y + 6 + y = 10$$
$$2y + 6 = 10$$

4. Say to them: "Now you can solve for *y*."

$$2y + 6 = 10$$

5. Say to your child: "Now that you know the value of *y*, which is 2, you can substitute it into one of the expressions to solve for *x*."

$$y + x = 10$$
$$2 + x = 10$$

6. Tell them: "To get *x* alone on one side of the equal sign (to isolate *x*), you subtract 2 from both sides."

$$-2 + 2 + x = 10 - 2$$
$$x = 8$$

7. Say to your child: "The value of *x* is 10 and the value of *y* is 2. If you plug both values into the expressions, you should get the following results."

$$y + x = 10 \qquad x - y = 6$$
$$2 + 8 = 10 \qquad 8 - 2 = 6$$
$$10 = 10 \qquad 6 = 6$$

Ask your child to consider the following expressions:

$$x - y = 8$$
$$x + y = 12$$

1. To start, use one of the expressions to solve for *x*:

$$x - y = 8$$

2. Tell your child: "Your goal is to get *x* by itself on one side of the equal sign. To do so, you can add *y* to both sides."

$$x - y + y = 8 + y$$
$$x = 8 + y$$

3. Say to your child: "Now that you know the value of *x*, you can substitute its value $(8 + y)$ into one of the expressions and solve for *y*."

$$x + y = 12$$
$$y + 8 + y = 12$$
$$2y + 8 = 12$$

4. **Say to them: "To isolate y on one side of the equal sign, you subtract 8 from both sides and then divide by 2."**

$$2y + 8 - 8 = 12 - 8$$
$$\frac{2y}{2} = \frac{4}{2}$$
$$y = 2$$

5. **Tell them: "Now that you know the value of y, which is 2, you can substitute it into one of the expressions to solve for x."**

$$x + y = 12$$
$$x + 2 = 12$$
$$x + 2 - 2 = 12 - 2$$
$$x = 10$$

6. **Say to your child: "The value of x is 10 and the value of y is 2. If you plug both values into the expressions, you should get the following results."**

$$x + y = 12 \qquad x - y = 8$$
$$10 + 2 = 12 \qquad 10 - 2 = 8$$
$$12 = 12 \qquad 8 = 8$$

Ask your child to solve the following expressions:

$$x - y = 3$$
$$x + y = 5$$

1. **To start, use one of the expressions to solve for x:**

$$x - y = 3$$

2. **Tell your child: "Your goal is to get x by itself on one side of the equal sign. To do so, you can add y to both sides."**

$$x - y + y = 3 + y$$
$$x = 3 + y$$

3. **Say to your child: "Now that you know the value of x, you can substitute its value (3 + y) into one of the expressions and solve for y."**

$$x + y = 5$$
$$3 + y + y = 5$$
$$2y + 3 = 5$$

4. **Say to them: "To isolate y on one side of the equal sign, you subtract 3 from both sides and then divide by 2."**

$$2y + 3 - 3 = 5 - 3$$
$$\frac{2y}{2} = \frac{2}{2}$$
$$y = 1$$

5. Tell them: "Now that you know the value of *y*, which is 1, you can substitute it into one of the expressions to solve for *x*."

$$x + y = 5$$
$$x + 1 = 5$$
$$x + 1 - 1 = 5 - 1$$
$$x = 4$$

6. Say to your child: "The value of *x* is 4 and the value of *y* is 1. If you plug both values into the expressions, you should get the following results."

$$x + y = 5 \qquad x - y = 3$$
$$4 + 1 = 5 \qquad 4 - 1 = 3$$
$$5 = 5 \qquad 3 = 3$$

Present the following expressions to your child:

$$x - 2y = 6$$
$$x + y = 12$$

1. To start, use one of the expressions to solve for *x*:

$$x + y = 12$$

2. Tell your child: "Your goal is to isolate *x* on one side of the equal sign. To do so, you can subtract *y* from both sides."

$$x + y - y = 12 - y$$
$$x = 12 - y$$

3. Say to your child: "Now that you know the value of *x*, you can substitute its value into one of the expressions and solve for *y*. To do this, replace the value of *x* with 12 – *y*."

$$x - 2y = 6$$
$$12 - y - 2y = 6$$
$$12 - 3y = 6$$

4. Say to them: "Now you need to isolate *y* on one side of the equal sign. To do so, you can add 3*y* to both sides."

$$12 + 3y - 3y = 6 + 3y$$
$$12 = 6 + 3y$$

5. Tell them: "Next, you can subtract 6 from both sides."

$$12 - 6 = 6 - 6 + 3y$$

6. Say to them: "Finally, you divide both sides by 3."

$$\frac{6}{3} = \frac{3y}{3}$$
$$2 = y$$

7. Explain to your child: "Now that you know the value of y, which is 2, you can substitute it into one of the expressions to solve for x."

$$x + y = 12$$
$$x + 2 = 12$$
$$x + 2 - 2 = 12 - 2$$
$$x = 10$$

8. Say to your child: "The value of x is 10 and the value of y is 2. If you plug both values into the expressions, you should get the following results."

$$x - 2y = 6 \qquad\qquad x + y = 12$$
$$10 - 4 = 6 \qquad\qquad 10 + 2 = 12$$
$$6 = 6 \qquad\qquad 12 = 12$$

FIND ONLINE

This book's companion website at www.dummies.com/go/teachingyourkids newmath6-8fd contains a worksheet (the first few rows of which are shown in Figure 23-2) that your child can use to solve for x and y. Download and print the worksheet. Help your child solve the first few and then ask them to complete the rest.

$$x + y = 5 \qquad\qquad\qquad\qquad x - y = 6$$

$$x - y = 3 \qquad\qquad\qquad\qquad x + y = 10$$

FIGURE 23-2:
A worksheet for using two expressions to solve for 2 unknowns.

$$x + y = 25 \qquad\qquad\qquad\qquad x - y = 0$$

$$y - x = 5 \qquad\qquad\qquad\qquad x + y = 6$$

Chapter **24**

Numbers, Like People, Aren't Always Rational

Your child knows how to work with a variety of numbers. They can count to over 1 million, and they have learned that numbers can be positive or negative. They know that decimal numbers have a decimal point. And they have learned to work with fractions.

In this chapter, your child reviews the number types. They should be able to describe a number by its type. They will learn that there are two broad classes of numbers:

» *Rational numbers,* which you can express in a finite number of digits to the right of the decimal point or for which the numbers to the right of decimal point repeat.

» *Irritational numbers* that don't end or repeat.

Then, to the joy of Buzz Lightyear, they will learn the meaning of "infinity and beyond!"

Reviewing Number Types

As your child studies math, they will encounter different number types, such as counting numbers, integers, and decimal numbers. Each number type has different characteristics that your child should know. In this section, you examine those number types.

Say to your child: "When you first learned to count numbers, you worked with counting numbers that normally start with 1 and increase indefinitely."

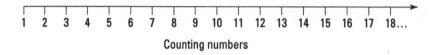

Counting numbers

Tell your child: "Counting numbers are called *rational numbers.*"

Explain to your child: "Counting numbers never end. Regardless of how big a number is, you can always add 1 to the number. Mathematicians use the word *infinity* and the symbol ∞ to represent the unending nature of such numbers."

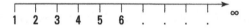

Say to your child: "In Chapter 10, you learned about negative numbers, which, like positive numbers, can go on forever, meaning, no matter how small a number is, you can always subtract 1 more. You call positive and negative whole numbers (numbers without a decimal point) *integers.*"

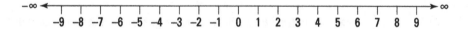

Tell your child: "A decimal number is a number with a decimal point, such as 3.1 or 5.612. As you learned in Chapter 7, some decimal numbers repeat."

$$\frac{1}{3} = 0.3\overline{3} \qquad \frac{2}{3} = 0.6\overline{6}$$

Explain to them: "You have worked with fractions such as $\frac{5}{1}$ or $\frac{3}{5}$. If you perform the fraction's division, your result is either an integer (such as $\frac{4}{4}$ equals 1) or a decimal number ($\frac{3}{5} = 0.6$)."

Getting a Handle on Rational and Irrational Numbers

In the previous section, your child reviewed the number types: counting, integer, decimal, and fractions. In this section, they learn about rational and irrational numbers. Like people, most numbers are rational, which means you can represent their value. Some numbers (like people) are irrational, which means they never stop!

Grasping the difference between rational and irrational numbers

Say to your child: "When mathematicians talk about numbers, they describe them as being either rational or irrational. In general, a rational number is a number whose value you can represent."

Here are some examples:

>> Integer numbers such as -5, 0, or 5

>> Decimal numbers such as 5.251

>> Decimal numbers that repeat, such as $0.3\overline{3}$

REMEMBER

If you're feeling more formal, here's the formal definition of rational numbers: A rational number is any number that can be expressed as a fraction $\frac{a}{b}$ where a and b are integers and b is not equal to 0.

When you're confident your child understands what a rational number is, you can explain irrational numbers. Say to them: "In contrast, an irrational number is a number you cannot represent because its value never ends and never repeats. Pi (π), for example, is an irrational number."

NEVER-ENDING PI

Pi (π) is an irrational number whose decimal values never end. Many people have tried to memorize digits of π. In fact, some people have memorized tens of thousands of its digits! Several sites on the web present the first million digits of π. Check out the sites with your child and challenge them to memorize the first 10!

Identifying rational and irrational numbers

Given the formal definition of a rational number, ask your child to consider the following numbers:

5	0.2	$0.3\bar{3}$

The number 5, like all integers, is rational because you can represent it using the fraction $\frac{5}{1}$.

The value 0.2 is rational because you can represent it using the fraction $\frac{1}{5}$.

The value $0.3\bar{3}$ is rational because you can represent it using the fraction $\frac{1}{3}$.

In addition to pi, many roots, such as $\sqrt{2}$ and $\sqrt{3}$, are irrational — their values never end.

Present the chart of number types to your child shown in Figure 24-1.

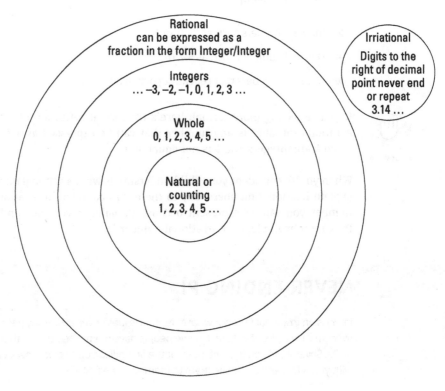

FIGURE 24-1:
Number types.

Chapter **25**

Raising the Status with Exponents

An *exponent* is the value or power to which you raise a number, such as the value 2 in 5^2. In previous chapters, your child learned to work with the power of 2. In this chapter, you raise the stakes and solve and chart expressions that have exponents:

$$y = x^2$$

Your child also learns to solve cube roots, such as $\sqrt[3]{27}$, which return the value that, when cubed, equals the number under the expression:

$\sqrt[3]{27}$

$3 \times 3 \times 3 = 27$

$\sqrt[3]{27} = 3$

Finally, your child learns to write decimal values such 123.456 in scientific notation (1.23456×10^2), which is a standard form for representing numbers used by scientists, statisticians, and other smart people.

So, let's get started — this chapter should be exponentially fun!

Solving Expressions with Exponents

In Chapter 14, you taught your child that an exponent is the power to which an expression is raised. In the case of 5^2, the exponent is 2. In Chapter 14, all the problems with exponents used only numbers, such as 3^2 or 4^2. In this section, your child uses exponents with expressions, such as:

$$(x-3)^2 = (x-3) \times (x-3)$$

Say to your child: "You have learned to solve problems that have exponents and numbers, like these."

$$2^2 = 4 \qquad 3^2 = 9 \qquad 2^4 = 16 \qquad 5^3 = 125$$

Tell them: "As you advance in math, you will encounter exponents with expressions, such as this."

$$(5-2)^2$$

Explain to your child: "You can represent such a squared expression as follows."

$$(5-2)^2 = (5-2) \times (5-2)$$

Say to them: "To solve this problem, you, as always, solve the expressions grouped within parentheses first."

$$(5-2)^2 = (5-2) \times (5-2)$$
$$= 3 \times 3$$
$$= 9$$

Explain to your child: "Often, you will encounter expressions that have a variable and an exponent."

$$(x+2)^2$$

Say to them: "Again, you can represent the expression like this."

$$(x+2)^2 = (x+2) \times (x+2)$$

Say to your child: "Assume you want to solve the expression for x with the value 5. You can solve the expression as follows."

$$
\begin{aligned}
(x+2)^2 &= (x+2) \times (x+2) \\
&= (5+2) \times (5+2) \\
&= 7 \times 7 \\
&= 49
\end{aligned}
$$

Charting Expressions

Your child has learned to work with expressions in the form:

$$y = x^2$$

In this section, they learn to chart such expressions which they will often do when they work with data such as ages and heights. In this case, the data are x and y, which they will plot using the corresponding chart axes.

Say to your child: "Consider the following expression."

$$y = x^2$$

Say to them: "Let's assume that you have the following data values for x."

x
1
2
3
4
5

Tell them: "Using the expression $y = x^2$ and the given values for x, calculate the corresponding values for y. In other words, knowing the value of x in the expression $y = x^2$, you will solve for y."

x	y
1	1
2	4
3	9
4	16
5	25

Ask your child to use the following graph to chart the data:

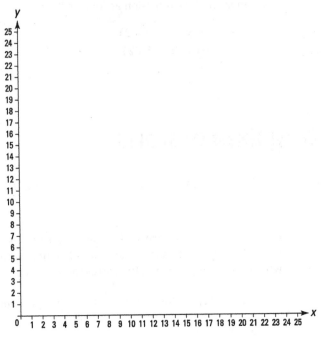

Your child should get:

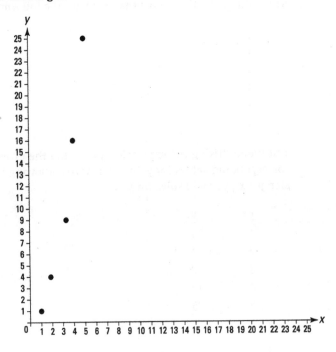

Ask your child to consider the following expression:

$$y = (x-1)^2$$

1. Say to your child: "Suppose you are given the following data for x."

x
1
2
3
4
5

2. Ask your child to use the values of x and the expression to solve for y. They should get:

x	y
1	0
2	1
3	4
4	9
5	16

If they do not, review the following with them:

x	
1	$y = (1-1)^2 = (0)^2 = 0$
2	$y = (2-1)^2 = (1)^2 = 1$
3	$y = (3-1)^2 = (2)^2 = 4$
4	$y = (4-1)^2 = (3)^2 = 9$
5	$y = (5-1)^2 = (4)^2 = 16$

3. **Ask your child to use the following graph to plot the data values:**

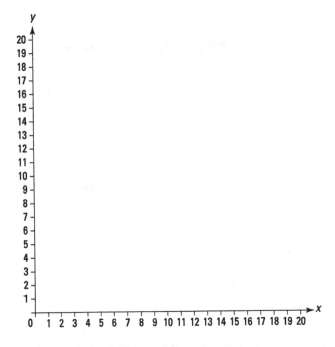

Your child should get the following:

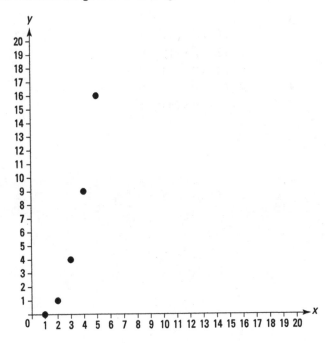

Calculating Cube Roots

In Chapter 14, you taught your child about the square root that determines the value that, when squared, equals the number specified under the square-root symbol:

$$\sqrt{4} = 2 \qquad \sqrt{9} = 3 \qquad \sqrt{16} = 4$$

In this section, you teach your child about the *cube root*, $\sqrt[3]{}$, which determines the value that, when cubed (raised to the power of 3), equals the number specified under the cube-root symbol:

$$\sqrt[3]{27} = 3 \qquad \sqrt[3]{125} = 5 \qquad \sqrt[3]{64} = 4$$

Say to your child: "You have learned to solve for the square root, $\sqrt{}$, which determines the value that when squared, equals the number under the square-root symbol."

$$\sqrt{4} = 2 \qquad \sqrt{9} = 3 \qquad \sqrt{25} = 5$$

Say to them: "You have also learned that you can cube a number by raising the number to a power of 3."

$$2^3 = 8 \qquad 3^3 = 27 \qquad 4^3 = 64 \qquad 10^3 = 1,000$$

Explain to them: "Now you will learn about the cube root, $\sqrt[3]{}$, which returns the value that, when cubed, equals the number specified under the cube-root symbol."

$$\sqrt[3]{8} = 2 \qquad \sqrt[3]{27} = 3 \qquad \sqrt[3]{64} = 4$$

Explain to your child: "Solving the cube root can be difficult — the previous examples were easy! You may want to use a calculator to solve for the cube root so you can get exact results. Many calculators have square-root and cube-root buttons to help you calculate the square root of a value."

Say to them: "That said, you can estimate the cube root for different values. Assume, for example, that you have the following."

$$\sqrt[3]{35}$$

Say to your child: "To start, you might guess 3."

$$3 \times 3 \times 3 = 27$$

Tell them: "Then, you might try 4."

$$4 \times 4 \times 4 = 64$$

Explain to your child: "Since the number 35 falls between 27 and 64, you know the cube root of 35 is 3 and some decimal values. You might guess 3.2, for example."

$3.2 \times 3.2 \times 3.2 = 35.968$ That's pretty close!

Understanding Negative Exponents

Your child has learned to work with numbers that have exponents, such as:

$$5^2 \qquad 2^3 \qquad 4^5 \qquad 6^4$$

Your child has learned that a positive exponent tells them how many times they should multiply a value times itself:

$$5^2 = 5 \times 5$$
$$2^3 = 2 \times 2 \times 2$$
$$4^5 = 4 \times 4 \times 4 \times 4 \times 4$$
$$6^4 = 6 \times 6 \times 6 \times 6$$

Just as you can have positive exponents, you can also have negative exponents:

$$5^{-1} \qquad 5^{-2} \qquad 5^{-3}$$

A negative exponent tells you how many times you must multiply the reciprocal of the value (the reciprocal of 5 is $\frac{1}{5}$) times itself:

$$5^{-1} = \frac{1}{5}$$
$$5^{-2} = \frac{1}{5^2} = \frac{1}{25}$$
$$5^{-3} = \frac{1}{5^3} = \frac{1}{125}$$

Say to your child: "You know how to solve values raised to a positive exponent, such as these."

$$3^2 \qquad 3^3 \qquad 3^4$$

Explain to them: "The positive exponent tells you how many times you multiply the value times itself."

$$3^2 = 3 \times 3$$
$$3^3 = 3 \times 3 \times 3$$
$$3^4 = 3 \times 3 \times 3 \times 3$$

Say to your child: "Just as you can have positive exponents, you can also have negative exponents."

$$3^{-2} \qquad 3^{-3} \qquad 3^{-4}$$

Explain to them: "A negative exponent tells you how many times you multiply the reciprocal of the value times itself. The reciprocal of 3 is $\frac{1}{3}$. For the previous numbers you get the following."

$$3^{-2} = \frac{1}{3^2} = \frac{1}{3} \times \frac{1}{3} = \frac{1}{9}$$

$$3^{-3} = \frac{1}{3^3} = \frac{1}{3} \times \frac{1}{3} \times \frac{1}{3} = \frac{1}{27}$$

$$3^{-4} = \frac{1}{3^4} = \frac{1}{3} \times \frac{1}{3} \times \frac{1}{3} \times \frac{1}{3} = \frac{1}{81}$$

Ask your child to write the following values and exponents in their reciprocal form:

$$3^{-2} \qquad 7^{-3} \qquad 8^{-4}$$

They should get:

$$\frac{1}{3^2} \qquad \frac{1}{7^3} \qquad \frac{1}{8^4}$$

In a similar way, ask your child to represent the following values in a value raised to negative exponent:

$$\frac{1}{5^3} \qquad \frac{1}{6^2} \qquad \frac{1}{9^3}$$

They should get:

$$5^{-3} \qquad 6^{-2} \qquad 9^{-3}$$

Multiplying Numbers with Exponents

As your child encounters more complex math problems, there will be times when they must multiply values that have exponents, such as:

$$5^2 \times 5^2 = \qquad 7^2 \times 7^3 = \qquad 6^2 \times 6^4 =$$

When such multiplication numbers have the same base value, such as $5^2 \times 5^2$, you solve the problem by adding the exponents:

$$5^2 \times 5^2 = 5^4 \qquad 7^2 \times 7^3 = 7^5 \qquad 6^2 \times 6^4 = 6^6$$

Say to your child: "As you encounter expressions that use exponents, there may be times when you must multiply the same number."

$$2^2 \times 2^3 = \qquad 3^3 \times 3^4 = \qquad 4^2 \times 4^5 =$$

Explain to them: "To solve these problems (remember, the base number must be the same), you add the exponents."

$$2^2 \times 2^3 = 2^5 \qquad 3^3 \times 3^4 = 3^7 \qquad 4^2 \times 4^5 = 4^7$$

Ask your child to solve the following expressions:

$$3^2 \times 3^4 = \qquad 4^2 \times 4^3 = \qquad 5^2 \times 5^2 =$$

They should get:

$$3^2 \times 3^4 = 3^6 \qquad 4^2 \times 4^3 = 4^5 \qquad 5^2 \times 5^2 = 5^4$$

Explain to your child: "There may be times when one or both of the numbers have negative exponents."

$$3^4 \times 3^{-2} = \qquad 2^5 \times 2^{-3} = \qquad 4^4 \times 4^{-3} =$$

In such cases, you again add the exponents:

$$3^4 \times 3^{-2} = 3^2 \qquad 2^5 \times 2^{-3} = 2^2 \qquad 4^4 \times 4^{-3} = 4^1 = 4$$

Ask your child to solve the following expressions:

$$3^3 \times 3^{-2} = \qquad 4^{-2} \times 4^5 = \qquad 5^2 \times 5^{-3} =$$

They should get:

$$3^3 \times 3^{-2} = 3^1 = 3 \qquad 4^{-2} \times 4^5 = 4^3 \qquad 5^2 \times 5^{-3} = 5^{-1}$$

Dividing Numbers with Exponents

Just as there will be times when your child must solve expressions that multiply the same base number with exponents, there will also be times when they must divide them, as in these examples:

$$3^4 \div 3^2 = \qquad 2^5 \div 2^3 = \qquad 4^4 \div 4^3 =$$

In such cases, you subtract the exponents:

$$3^4 \div 3^2 = 3^2 \qquad\qquad 2^5 \div 2^3 = 2^2 \qquad\qquad 4^4 \div 4^3 = 4^1 = 4$$

Say to your child: "You have learned to multiply the same base number with exponents by adding the exponents. There will also be times when you must divide them."

$$2^4 \div 2^2 = \qquad\qquad 3^5 \div 3^3 = \qquad\qquad 5^4 \div 5^3 =$$

Say to them: "To divide such values, you subtract the exponents."

$$2^4 \div 2^2 = 2^2 \qquad\qquad 3^5 \div 3^3 = 3^2 \qquad\qquad 5^4 \div 5^3 = 5^1 = 5$$

Ask your child to divide the following values:

$$5^5 \div 5^3 = \qquad\qquad 4^5 \div 4^2 = \qquad\qquad 3^4 \div 3^3 =$$

They should get:

$$5^5 \div 5^3 = 5^2 \qquad\qquad 4^5 \div 4^2 = 4^3 \qquad\qquad 3^4 \div 3^3 = 3^1 = 1$$

Solving Problems with Exponents

Your child is now ready to solve addition and subtraction problems that contain exponents. Present the following problems to your child:

$$3^{-1} + 2^2 = \qquad 5^2 + 5^{-2} = \qquad 4^2 - \left(\frac{1}{3}\right)^{-2} =$$

Start with the problem $3^{-1} + 2^2$.

1. **Rewrite 3^{-1} as $\frac{1}{3}$:**

   $$\frac{1}{3} + 2^2 =$$

2. **Rewrite 2^2 as 4 and solve the problem:**

   $$\frac{1}{3} + 4 = 4\frac{1}{3}$$

Next, consider the expression $5^2 + 5^{-2}$.

1. **Rewrite 5^2 as 25:**

   $$25 + 5^{-2} =$$

2. Rewrite 5^{-2} as $\dfrac{1}{5^2}$ or $\dfrac{1}{25}$ and solve the problem:

$$25 + \frac{1}{25} = 25\frac{1}{25}$$

Finally, consider the expression $4^2 - \left(\dfrac{1}{3}\right)^{-2}$.

1. Rewrite 4^2 as 16:

$$16 - \left(\frac{1}{3}\right)^{-2} =$$

2. Rewrite $\left(\dfrac{1}{3}\right)^{-2}$ using the reciprocal of $\dfrac{1}{3}$, which is 3:

$$16 - 3^2 =$$

3. Rewrite 3^2 as 9 and solve the problem:

$$16 - 9 = 7$$

Introducing Scientific Notation

When you perform arithmetic operations, the results you get for different expressions may take the following forms:

$$120.6 \times 157.2 = 18958.32$$
$$25.5\overline{)1864} = 73.0980392$$
$$5.65^2 = 31.9225$$
$$1 - 0.28 = 0.72$$

Around the world, mathematicians (and scientists and physicists) often convert such values to a standard format called *scientific notation*, which lets you put the decimal point after the first non-zero number:

$$18958.32 = 1.895832 \times 10^4$$
$$46.435 = 4.6435 \times 10^1$$
$$31.9225 = 3.19225 \times 10^1$$
$$0.72 = 7.2 \times 10^{-1}$$

REMEMBER

To determine the exponent for the power of 10, you count the number of places you move the decimal point. If you move the decimal point to the left, the exponent is positive. Likewise, if you move the decimal point to the right, the exponent is negative.

$$303.735 = 3.03735 \times 10^2 \qquad\qquad 0.004563 = 4.563 \times 10^{-3}$$

Using scientific notation, not only do you put numbers in a consistent form, but someone looking at the exponent values can also tell quickly if the number is big or small.

The reason scientific notation uses the power of 10 is so that you can move the decimal, which is based on powers of 10.

In this section, you teach your child about scientific notation.

Say to your child: "Assume you solved different math problems and got the following results."

0.0235	100265.7	876.35	4437.375

Say to them: "If you look at the numbers, the decimal points appear at different locations."

Explain to your child: "When mathematicians, scientists, and physicists display the results of a problem, they often use a standard format called scientific notation."

Say to them: "Using scientific notation, you move the decimal point after the first non-zero number. Then, to tell others how many places you moved the decimal point, you multiply the value times 10 raised to the count of places moved. If you moved the decimal point left, the exponent is positive. If you moved the decimal point right, the exponent is negative."

Ask your child to consider the following:

123.789

Explain to your child: "To put the value into scientific notation, you must move the decimal point 2 places to the left. As such, you use the following form."

1.23789×10^2

Ask your child to consider the following number:

37425.683

Say to them: "In this case, to put the number into scientific notation, you must move the decimal point 4 places to the left, yielding the following result."

3.7425683×10^4

Say to your child: "Using scientific notation does not change the result, but rather, it puts the result in a standard format."

Ask your child to consider the following value:

0.0003421

Say to them: "In this case, you must move the decimal point 4 places to the right. As a result, you use a negative exponent."

3.421×10^{-4}

Ask your child to put the following values into scientific notation:

0.0235	100265.7	876.35	4437.375

They should get:

2.35×10^{-2} 1.002657×10^{5} 8.7635×10^{2} 4.437375×10^{3}

If they do not, help your child count the number of places the decimal point moves and use that value for an exponent.

FIND
ONLINE

This book's companion website at www.dummies.com/go/teachingyourkids newmath6–8fd contains a worksheet (the first few rows of which are shown in Figure 25-1) that your child can use to practice working with scientific notation. Download and print the worksheet. Help your child solve the first few and then ask them to complete the rest.

1234.5678   5555.221   12.34567

FIGURE 25-1:
Working with
scientific
notation.   133.2222   2212.333   11.23443

Rewriting Numbers from Scientific Notation

In the previous section, your child learned how to write numbers using scientific notation. In this section, they do just the opposite, rewriting numbers from scientific notation to their decimal form.

Say to your child: "Assume that you have the following number."

$$1.3754 \times 10^2$$

Ask your child to write the number in its original decimal form. Say to your child: "In this case, the exponent is 2 and specifies the number of places to the left the decimal point is moved. To write the number in its original form, you move the decimal point that many places to the right."

137.54

Ask your child to consider the following number:

$$1.321 \times 10^{-3}$$

Say to your child: "In this case, because the exponent is negative, you must move the decimal point to the left, adding zeros."

$$1.321 \times 10^{-3} = 0.001321$$

Ask your child to rewrite the following numbers in their decimal form:

$$3.630 \times 10^2 \qquad 5.514367 \times 10^5 \qquad 1.23 \times 10^{-2}$$

They should get:

| 363.0 | 551436.7 | 0.0123 |

If they do not, help your child to use the exponent to move the decimal point to the right or left as needed.

FIND ONLINE

This book's companion website at www.dummies.com/go/teachingyourkids newmath6–8fd contains a worksheet (the first few rows of which are shown in Figure 25-2) that your child can use to rewrite numbers from scientific notation to their decimal form. Download and print the worksheet. Help your child solve the first few and then ask them to solve the rest.

FIGURE 25-2: A worksheet for rewriting numbers from scientific notation.

| 1.2345678×10^5 | 5.555221×10^2 | 1.234567×10^3 |

| 1.332222×10^4 | 2.212333×10^2 | 1.123443×10^{-3} |

Chapter **26**

Calculating a Line's Slope and Intercept

ines. When you are standing in them, they get annoying. When you need to go from point a to point b, a straight line is the shortest distance. Lines are all around us. In this chapter, your child finds out how to work with and chart equations (called *linear equations*) that, when charted, produce a line.

So, get ready for more lines and let's get started!

Comparing Linear and Nonlinear Equations

Throughout this book, your child plots the data for different data sets. Consider the following *x* and *y* values that correspond to the equation $y = 2x + 1$:

x	y
0	1
1	3

x	y
2	5
3	7
4	9
5	11

Using a chart with *x* and *y*, you would plot the data points as shown here:

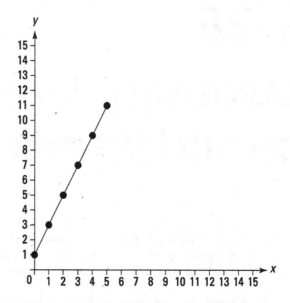

This chapter focuses on linear equations, which result in a line when you chart the data (hence the term *linear*).

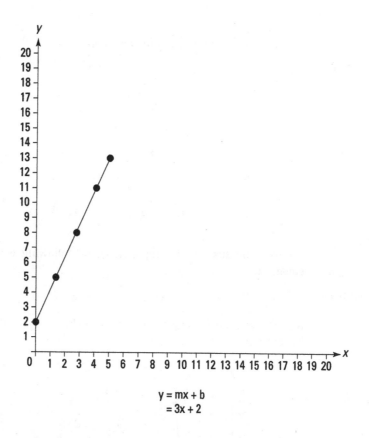

$$y = mx + b$$
$$= 3x + 2$$

REMEMBER

It turns out that linear expressions do not have a squared (or greater) exponent. For example, the following expressions are linear equations:

$$y = 3x \qquad y = 5x - 1 \qquad y = 7x - 2$$

Say you wanted to graph each of these expressions for the following data set:

x
1
2
3
4
5

You would get the following linear graphs:

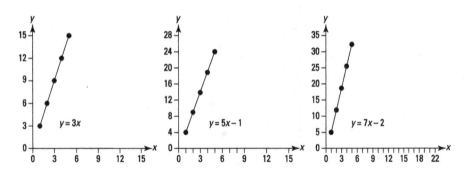

REMEMBER

Likewise, because of the exponents, the following expressions are not linear equations:

$$y = x^2 \qquad y = x^3 - 4 \qquad y = x^4 - 7$$

Again, if you were to plot the previous x values, you would get the following non-linear graphs:

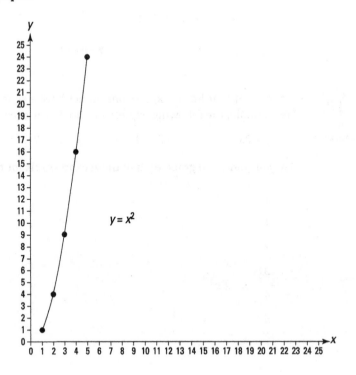

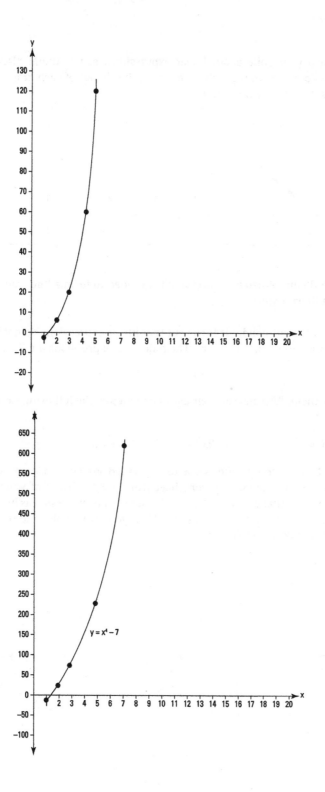

$y = x^4 - 7$

To teach your child about linear expressions, say to them: "Depending on the expressions that your graph is based on, the shapes of each graph, as you connect points with a line, will differ."

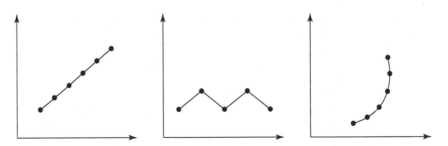

Say to them: "When an expression is graphed to form a line, you call the expression a linear expression."

Explain to your child: "It turns out that linear expressions do not have exponents of 2 or greater, such as $y = x^3$. The following expressions are linear."

$$y = x \qquad\qquad y = x + 1 \qquad\qquad y = 2x - 3$$

Say to them: "Because of their use of exponents, the following expressions are not linear."

$$y = x^2 \qquad\qquad y = 2x^3 - 1 \qquad\qquad y = x^3$$

This book's companion website at www.dummies.com/go/teachingyourkids newmath6–8fd contains a worksheet (the first few rows of which are shown in Figure 26-1) that your child can use to identify an expression as linear or nonlinear. Download and print the worksheet. Help your child solve the first few problems and then ask them to complete the rest.

$y = 3x + 1$ $\qquad\qquad\qquad$ $y = 4x + 0$ $\qquad\qquad\qquad$ $y = 5x$

FIGURE 26-1:
A worksheet to identify linear expressions. $\quad y = 3x^2$ $\qquad\qquad\qquad$ $y = x$ $\qquad\qquad\qquad$ $y = 2x + 8$

Understanding the Slope and Intercept of a Linear Equation

When you look at a graph of a linear equation, the line may go up, down, or across the chart:

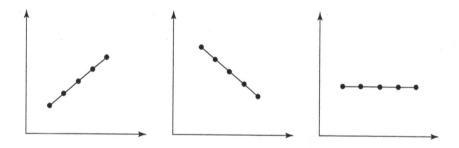

The slope of a graph tells you the direction the graph is headed as *x* values increase. In this section, you teach your child how to calculate a line's slope.

Say to your child: "When you graph a linear equation, you use the term *slope* to describe the direction of the line. As *x* values increase, a positive slope goes up and a negative slope goes down."

Explain to them: "The slope of a line is a number that indicates how far the line moves up (or down) as it moves across the page. The following graphs have lines with different slopes."

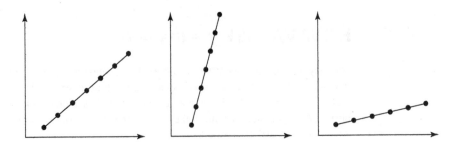

Say to your child: "Mathematicians describe the process of calculating the slope as solving the rise of the line (up or down) over the run of the line (across the page)."

$$\text{Slope} = \frac{\text{rise}}{\text{run}}$$

1. Tell your child: "To calculate the slope of a line, pick two points on the line and draw a vertical dashed line, as shown here."

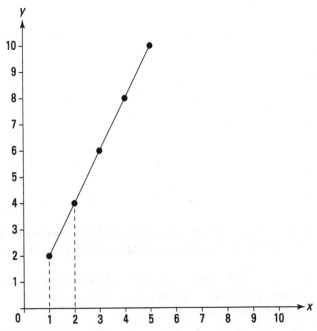

2. Say to them: "Next, pick two points on the y-axis and draw a horizontal dashed line that intersects your other line, as shown here."

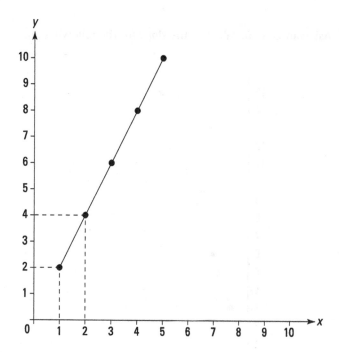

3. Explain to your child: "To determine the slope, you calculate the rise (change in y) over the run (change in x). To calculate the rise, you subtract the smaller y value from the larger one ($y_2 - y_1$)."

$$y_2 - y_1 =$$
$$4 - 2 = 2$$

4. Say to them: "To calculate the run, you subtract the smaller x value from the larger one ($x_2 - x_1$)."

$$x_2 - x_1 =$$
$$2 - 1 = 1$$

5. Say to your child: "Finally, you divide the rise by the run."

$$\frac{(y_2 - y_1)}{(x_2 - x_1)} = \frac{2}{1} = 2$$

6. Tell your child: "In this case, the line's slope is 2."

Ask your child to calculate the slope for the following line:

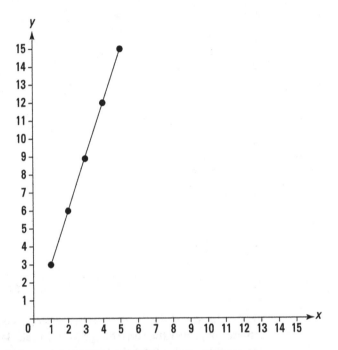

1. Say to your child: "To calculate the slope of a line, pick two points on the line and draw a vertical dashed line, as shown here."

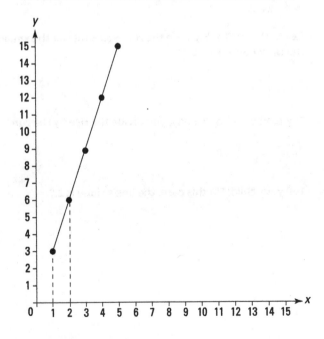

2. Tell them: "Next, pick two points on the *y*-axis and draw a horizontal dashed line that intersects your other line, as shown here."

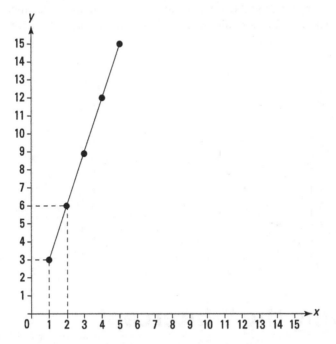

3. Explain to your child: "To determine the slope, you calculate the rise (change in *y*) over the run (change in *x*). To calculate the rise, you subtract the smaller *y* value from the larger one ($y_2 - y_1$)."

$$(y_2 - y_1) =$$
$$6 - 3 = 3$$

4. Say to them: "To calculate the run, you subtract the smaller *x* value from the larger one ($x_2 - x_1$)."

$$(x_2 - x_1) =$$
$$2 - 1 = 1$$

5. Say to your child: "Finally, you divide the rise by the run."

$$\frac{3}{1} = 3$$

6. Say to them: "In this case, the line's slope is 3."

FIND ONLINE

This book's companion website at www.dummies.com/go/teachingyourkidsnew math6-8fd contains a worksheet (the first part of which is shown in Figure 26-2) that your child can use to calculate the slope of lines. Download and print the worksheet. Help your child with the first few and then ask them to complete the rest.

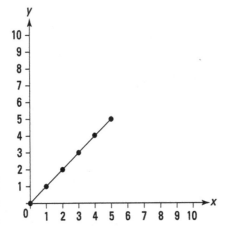
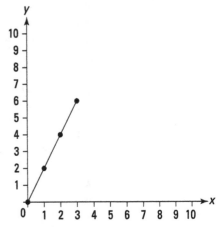

FIGURE 26-2:
A worksheet for calculating the slope of lines.

Identifying the Y-Intercept

To describe a line, your child will ultimately specify not only the line's slope, but also the line's y-intercept (the value of y where the line crosses the y-axis). The line will always cross the y-axis at the point where $x = 0$. Consider the following lines and y-intercept values:

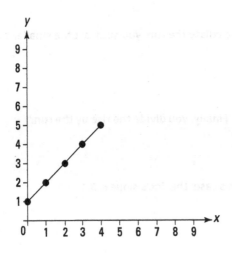
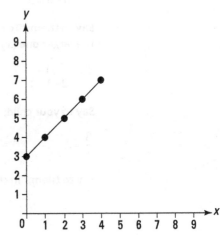

In this section, you help your child learn to identify the y-intercept.

Say to your child: "To specify a linear equation (an equation that results in a line), you must know two things: the line's slope and the value of y at the point where the line crosses the y-axis. You call that value the y-intercept."

Ask your child to consider the following graphs and to note the value of y at the point where the line crosses the y-axis:

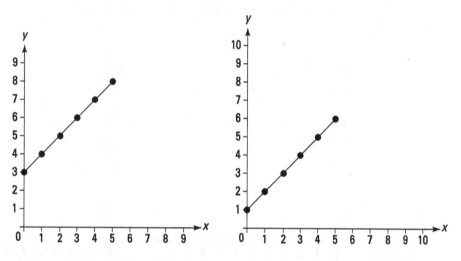

Ask your child to identify the y-intercept for the following lines:

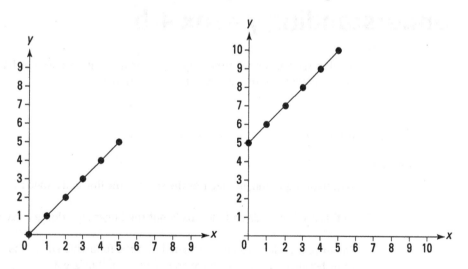

They should get:

y-intercept = 0   y-intercept = 5

If they don't, show your child the point where the line crosses the y-axis and note the corresponding y value.

**FIND
ONLINE**

This book's companion website at www.dummies.com/go/teachingyourkidsnew math6–8fd contains a worksheet (the first part of which is shown in Figure 26-3) that your child can use to identify the y-intercept.

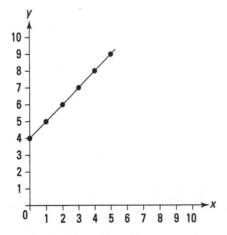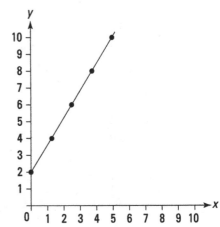

FIGURE 26-3:
A worksheet for
identifying the
y-intercept.

Understanding y = mx + b

As discussed earlier in this chapter, a linear equation does not have an exponent of 2 or greater. Here are some linear equations:

$$y = 3x \qquad y = 4x + 1 \qquad y = 2x - 2$$

REMEMBER

It turns out that there is a standard form for a linear expression:

$$y = mx + b$$

Within the equation, m specifies the slope of the line and b specifies the y-intercept.

In this section, your child learns about the linear equation $y = mx + b$.

Say to your child: "You have learned that the following equations are linear equations because they do not have an exponent of 2 or greater."

$$y = 4x \qquad y = 2x + 5 \qquad y = 4x - 2$$

Say to them: "It turns out that all linear equations follow this standard format."

$$y = mx + b$$

Explain to your child: "The letter m corresponds to the line's slope, and the b specifies the line's y-intercept."

Ask your child to consider the following expression:

$$y = 3x + 2$$

Explain to them: "In this case, the 3 in 3x tells you that the line's slope is 3. Likewise, the 2 represents the line's y-intercept."

$$y = \underset{\text{slope}}{m} \ x \ + \ \underset{\text{y–intercept}}{b}$$
$$y = \underset{\text{slope}}{3} \ x \ + \ \underset{\text{y–intercept}}{2}$$

Ask your child to identify the slope and intercept for the following linear equations:

$$y = 4x + 3 \qquad y = -2x - 5 \qquad y = 2x$$

They should get:

Slope: 4	Slope: –2	Slope: 2
y-intercept: 3	y-intercept: –5	y-intercept: 0

If they do not, review the equation $y = mx + b$ with your child and help them identify the corresponding slope and y-intercept.

FIND ONLINE

This book's companion website at www.dummies.com/go/teachingyourkids newmath6-8fd contains a worksheet (the first few rows of which are shown in Figure 26-4) that your child can use to identify the slope and intercept for linear equations. Download and print the worksheet. Help your child solve the first few and then ask them to complete the rest.

$y = 3x + 1$

Slope:
Y-intercept:

$y = 4x + 0$

Slope:
Y-intercept:

$y = 5x$

Slope:
Y-intercept:

$y = 3x$

Slope:
Y-intercept:

$y = x$

Slope:
Y-intercept:

$y = 2x + 8$

Slope:
Y-intercept:

FIGURE 26-4:
A worksheet for identifying the slope and intercept.

IN THIS CHAPTER

» Calculating perimeters for
uncommon shapes

» Finding areas for uncommon shapes

» Using the Pythagorean Theorem

» Determining the volume for three-
dimensional shapes

Chapter **27**

Circling Back to Geometry

Geometry is the branch of math that examines angles and shapes. In previous chapters, your child learned to calculate the perimeter and area of common shapes, such as squares, circles, and rectangles. In this chapter, they do the same, but for less common shapes such as trapezoids and parallelograms. If you don't know those two shapes, relax; I discuss them here.

Then, because real-world shapes are three-dimensional, you look at some 3D shapes and learn to calculate their volume — you know, how much stuff they could hold.

So, let's get in shape and get started!

Revisiting Perimeters

In Chapter 9, your child learned to calculate the perimeter, or distance around, common shapes. In this section, your child again calculates perimeters. However, this time, the shapes aren't so common.

Say to your child: "Previously, you learned to calculate the perimeter of common shapes, such as a rectangle, square, triangle, and circle. You will now perform similar processes, but this time for less common shapes."

Present the following parallelogram to your child:

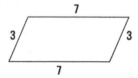

Explain to your child: "A *parallelogram* is a four-sided shape for which the opposite sides of the shape are parallel."

Say to them: "To determine the perimeter of the parallelogram, you add up the lengths of each side."

In this case, the perimeter is:

$3 + 7 + 3 + 7 = 20$

Present the following trapezoid to your child:

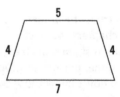

Explain to your child: "A *trapezoid* is a four-sided shape with only one set of parallel sides. To calculate the perimeter of a trapezoid, you add up the lengths of each side."

$4 + 5 + 7 + 4$

Present the following rectangle to your child:

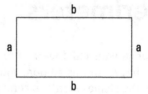

Explain to your child: "When you perform algebra, you often use letters to represent variables and unknowns."

Ask your child to calculate the rectangle's perimeter. They should get:

$a+b+a+b=2a+2b$

Ask your child to solve the rectangle's perimeter if a = 5 and b = 10:

$a+b+a+b=$
$5+10+5+10=30$

Present the following shape to your child:

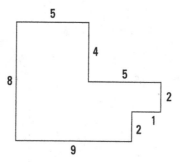

Ask your child to calculate the shape's perimeter. They should get:

$8+5+4+5+2+1+2+9=36$

If they do not, help your child add the length of each side of the shape.

Present the following shape to your child:

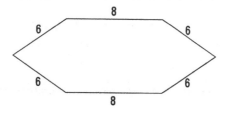

Ask your child to calculate the shape's perimeter. They should get:

$6+6+8+6+6+8=40$

FIND ONLINE

This book's companion website at www.dummies.com/go/teachingyourkidsnew math6–8fd contains a worksheet (the first part of which is shown in Figure 27-1) that your child can use to calculate the perimeter of shapes. Download and print the worksheet. Help your child solve the first few and then ask them to complete the rest.

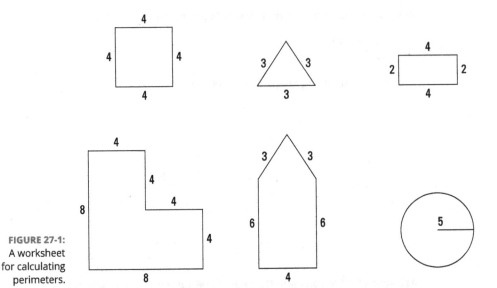

FIGURE 27-1: A worksheet for calculating perimeters.

Revisiting Areas

In Chapter 9, your child learned to calculate the area for common shapes. In this section, they calculate the area inside some not-so-common shapes.

Say to your child: "You have learned how to calculate the area for shapes such as rectangles, squares, triangles, and circles. Now, you will learn to calculate the area for some less common shapes."

Present the following parallelogram to your child:

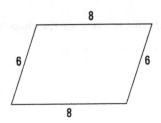

Explain to your child: "To calculate the area of a parallelogram, you use the same equation that you use to calculate the area of a rectangle: area = length × width."

Ask your child to calculate the area. They should get:

$8 \times 6 = 48$

Present the following trapezoid to your child:

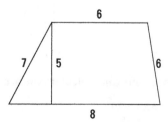

Say to your child: "To find the area of a trapezoid, you use the following equation."

$$\text{Area} = \frac{1}{2} \times (a+b) \times \text{height}$$

where a and b are the lengths of the parallel sides.

Your child should get:

$$\begin{aligned}
\text{Area} &= \frac{1}{2} \times (a+b) \times \text{height} \\
&= \frac{1}{2} \times (8+6) \times 5 \\
&= \frac{1}{2} \times (14) \times 5 \\
&= 7 \times 5 \\
&= 35
\end{aligned}$$

Present the following shape to your child:

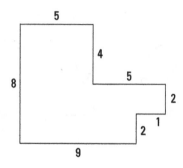

1. Ask your child to determine the shape's area.

2. Have your child divide the shape into 3 smaller shapes:

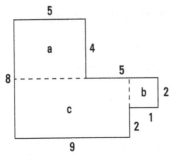

3. Have your child calculate the area for each of the smaller shapes:

$a = 5 \times 4 = 20$
$b = 2 \times 1 = 2$
$c = 9 \times 4 = 36$

4. Have your child add up the three areas:

$20 + 2 + 36 = 58$

FIND
ONLINE

This book's companion website at www.dummies.com/go/teachingyourkidsnew
math6-8fd contains a worksheet (the first part of which is shown in Figure 27-2)
that your child can use to calculate the area for various shapes. Download and
print the worksheet. Help your child solve the first few and then ask them to solve
the rest.

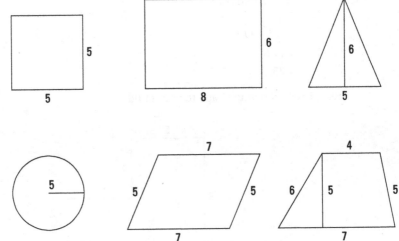

FIGURE 27-2:
A worksheet for
calculating the
area of shapes.

Understanding the Pythagorean Theorem

When your child works with right triangles (triangles with a 90° angle), there will be times when they don't know the length of one of the triangle's sides:

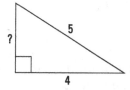

In this section, they learn to use the Pythagorean Theorem to solve the missing length:

$$c = \sqrt{a^2 + b^2}$$

Present the following right triangle to your child:

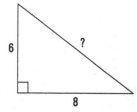

Say to your child: "This triangle is a right triangle, meaning it has a 90° angle, which is represented by the small box."

Explain to them: "In this case, you don't know the length of one of the triangle's sides — the hypotenuse, or long side."

Say to them: "When you have a right triangle, you can use the Pythagorean Theorem."

$$c = \sqrt{a^2 + b^2}$$

Say to your child: "In this case, you can calculate the length of the hypotenuse as follows."

$$c = \sqrt{a^2 + b^2}$$
$$= \sqrt{6^2 + 8^2}$$
$$= \sqrt{36 + 64}$$
$$= \sqrt{100}$$
$$= 10$$

Ask your child to use the Pythagorean Theorem to solve for the hypotenuse length of the following triangle:

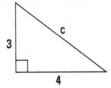

They should get:

$$c = \sqrt{a^2 + b^2}$$
$$= \sqrt{3^2 + 4^2}$$
$$= \sqrt{9 + 16}$$
$$= \sqrt{25}$$
$$= 5$$

If they don't, review the equation for the Pythagorean Theorem and help them solve the math.

Ask your child to determine the hypotenuse length (the length of the longest line, which is opposite of the right angle) for the following triangle:

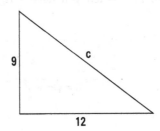

They should get:

$$c = \sqrt{a^2 + b^2}$$
$$= \sqrt{9^2 + 12^2}$$
$$= \sqrt{81 + 144}$$
$$= \sqrt{225}$$
$$= 15$$

Calculating a 3D Shape's Volume

Throughout this chapter, your child has calculated the area for a variety of shapes. When you work with a three-dimensional shape, such as a cube, sphere, or cylinder, you must calculate the shape's volume (the space inside the shape). In this section, your child calculates the volumes for common 3D shapes.

Present the following shapes to your child:

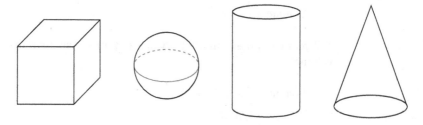

Explain to your child: "When you must know the space inside of a three-dimensional shape, you calculate the volume. In this section, you will learn the formulas to calculate the volume for each of these shapes."

Allow your child to use a calculator to perform the volume calculations.

TIP

Present the following cube to your child:

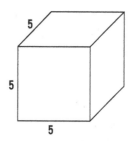

Say to your child: "To calculate the volume of a cube, you use this formula."

$$\text{Volume} = \text{length} \times \text{height} \times \text{width}$$

Ask your child to calculate the cube's volume. They should get:

$$\text{Volume} = 5 \times 5 \times 5$$
$$= 125$$

Present the following sphere to your child:

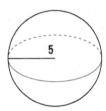

Say to them: "To calculate the volume for a sphere, you use the following formula."

$$\text{Volume} = \frac{4}{3} \times \pi \times r^3$$

Tell your child they can use the value 3.14 for π. Ask them to calculate the sphere's volume:

$$\text{Volume} = \frac{4}{3} \times 3.14 \times 5^3$$
$$= 523.333$$

Present the following cylinder to your child:

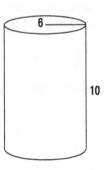

Say to your child: "To calculate the volume of a cylinder, you use the following formula."

$$\text{Volume} = \pi \times r^2 \times \text{height}$$

Ask them to calculate the cylinder's volume. They should get:

$$\text{Volume} = 3.14 \times 6^2 \times 10$$
$$= 1,130.4$$

Present the following cone to your child:

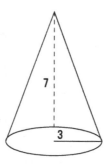

Tell your child: "To calculate the volume for a cone, you use the following formula."

$$\text{Volume} = \frac{1}{3}\pi \times r^2 \times \text{height}$$

Ask your child to calculate the cone's volume. They should get:

$$\text{Volume} = \frac{1}{3} \times 3.14 \times 3^2 \times 7$$
$$= 65.94$$

FIND ONLINE

This book's companion website at www.dummies.com/go/teachingyourkidsnew math6-8fd contains a worksheet (the first part of which is shown in Figure 27-3) that your child can use to find the volume for various 3D shapes. Download and print the worksheet. Help your child solve the first few and then ask them to solve the rest.

WHAT SIZE CONTAINER DO YOU NEED?

When you calculate the volume for a cube, you use the equation
volume = length × width × height. If your dimension measures are in feet, such as a
1-foot-by-2-foot-by-3-foot box, the volume is expressed in cubic feet: 6 cubic feet. It
turns out that 1 cubic foot can hold 7.48 gallons. So, a six-cubic-foot container can store
44.88 gallons. By the way, water weighs 8 pounds per gallon, so the full container would
weigh 359.04 pounds!

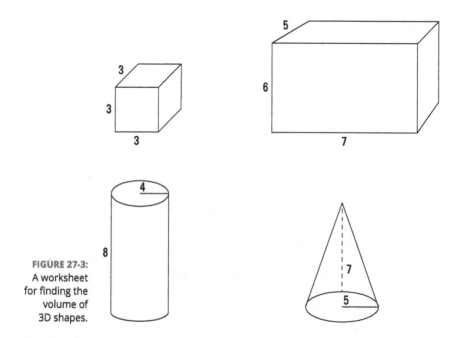

Working Out Geometric Word Problems

Okay, I'm guessing that after finding the volumes for various 3D shapes, word
problems don't seem so bad! In this section, your child solves word problems that
require them to find perimeters, areas, and volumes.

Present the following word problem to your child:

> Bill and Mary are putting grass in their backyard. Mary wants to put a plastic border
> around the grass and Bill wants to put sod.

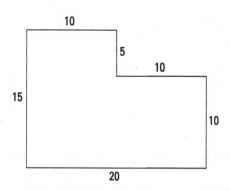

How much plastic must Mary buy? How much sod will Bill need?

1. **Calculate the perimeter of the yard by summing up the length of each side:**

   $15 + 10 + 5 + 10 + 10 + 20 = 60$

2. **To solve the area, you can divide the yard into two parts, which will create shapes for which you know how to calculate the area. Then you can calculate the area of each part:**

3.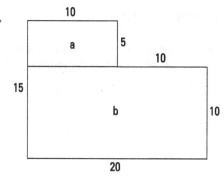

4. **Add up the two areas.**

   $15 \times 10 = 150$
   $10 \times 10 = 100$
   $150 + 100 = 250$

Present the following word problem to your child:

> Javiar sees how nice Bill and Mary's yard looks with grass and decides to put sod in his yard.

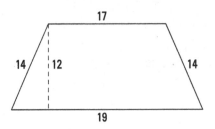

How much plastic border will Javiar need to buy? How much sod will he need?

1. **Calculate the perimeter by adding up the lengths of each side.**

   $14 + 17 + 14 + 19 = 64$

2. **Use the following area formula:**

   $$\text{Area} = \frac{1}{2} \times (a+b) \times \text{height}$$
   $$= \frac{1}{2} \times (17+19) \times 12$$
   $$= \frac{1}{2} \times 36 \times 12$$
   $$= 216$$

Present the following problem to your child:

Billy found two buckets in his garage. He wants to know which bucket can carry more water — meaning, he needs to know the volume of each bucket.

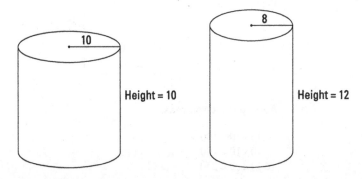

1. **Use the following equation to calculate each bucket's volume:**
   volume $= \pi \times \text{radius}^2 \times \text{height}$.

   Bucket 1 Volume
   $3.14 \times 10^2 \times 10 = 3,140$

   Bucket 2 Volume
   $3.14 \times 8^2 \times 12 = 2,411$

2. **Tell Billy the first bucket can store more water.**

Chapter **28**

Taking Another Chance on Probability and Statistics

Statistics is the branch of math that examines data. Your child has learned to perform simple statistics, such as finding the average, minimum, maximum, median, and modal values. In addition, your child has learned how to use variance to gain insights into data.

In this chapter, your child learns how to use charts to determine whether two variables, such as age and time spent playing video games, have a correlation. They also learn the importance of sample size in data analysis. Then, they learn how to apply a line of best fit to a data chart to gain insights.

Finally, for fun, your child learns how to solve *that* word problem in which two trains leave different train stations!

Creating Scatter Plots

A *scatter plot* is a graph of two sets of data, such as x and y data. In this section, you teach your child how to graph this data. Then, your child learns how to tell if there is a correlation between the data values.

Say to your child: "You have learned to work with x and y data values and expressions such as these."

$$y = x^2 - 1 \qquad y = x + 1$$

Consider the following data set for the expression $y = x^2 + 2$:

x	y
1	3
2	4
3	11
4	18
5	27

Say to your child: "Using a graph, you can plot the data as follows."

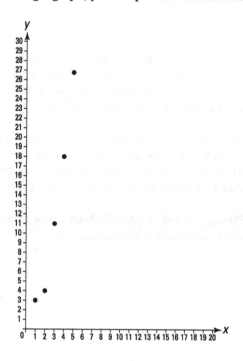

Tell your child: "To plot a point, you move across the *x*-axis to the corresponding value, and then you move up (or down) the *y*-axis to the corresponding value and plot the point."

Using the following chart, ask your child to plot the data points:

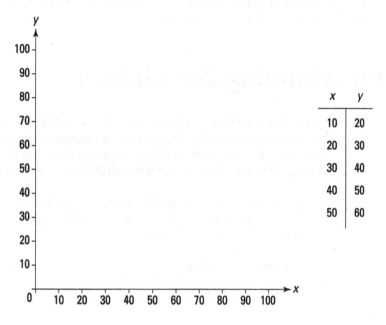

x	y
10	20
20	30
30	40
40	50
50	60

They should get:

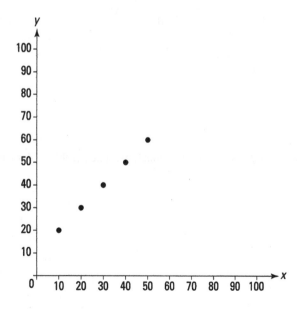

TIP

If they don't, help your child move across the x-axis and up the y-axis to the corresponding point.

FIND ONLINE

This book's companion website at www.dummies.com/go/teachingyourkidsnewmath6-8fd contains a worksheet that your child can use to plot data points. Download and print the worksheet. Help your child plot the first few data points and then ask them to complete the rest.

Understanding Correlation

Correlation is a fancy statistical term that describes whether a relationship exists between two variables. You might, for example, wonder if a correlation exists between shoe size and height, or a person's age and the amount of time they sleep. In this section, you teach your child how to identify such correlations.

Say to your child: "So far, you have used scatter plots to graph x and y data. Often, you will be given specific data to chart. Consider, for example, the following data that shows shoe size and height."

Shoe Size	Height
5	62
6	64
7	67
8	68
9	70
10	71
11	72
12	74

Using the following chart, ask your child to plot the data.

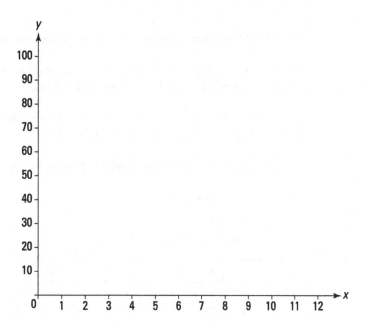

They should get:

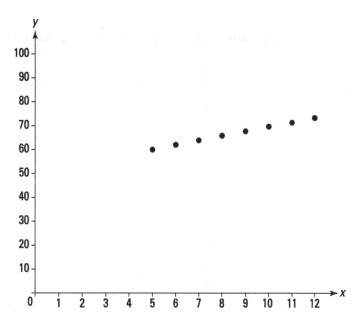

Ask your child: "By looking at the chart, what can you tell me about the relationship between shoe size and height? Meaning, as the shoe size gets bigger, what happens to the height?"

They should say that, as the shoe size gets bigger, the height also gets bigger.

Explain to your child: "Statisticians use the term *correlation* to describe whether or not there is a relationship between the data points."

Say to them: "In this case, you know that as the shoe size gets bigger, so too does the height. Statisticians call such as relationship a *positive correlation*."

Ask your child to consider the following data that compares age to hours slept.

Age	Hours Slept
1	12
5	10
10	8
21	7
50	6
70	5

Using the following chart, ask your child to plot the data:

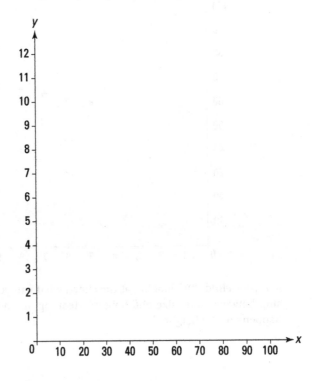

They should get:

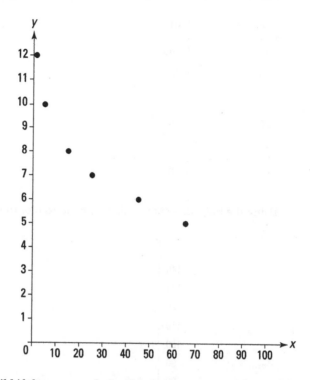

Ask your child if they see a relationship between age and hours slept — meaning, as age increases, what happens to the number of hours slept?

They should say that as age increases, the number of hours slept goes down.

Say to your child: "When you increase one value and the related value goes down, statisticians call that a *negative correlation*. If you look at the previous graph, you can see that the data is going down, which means age and hours slept have a negative correlation."

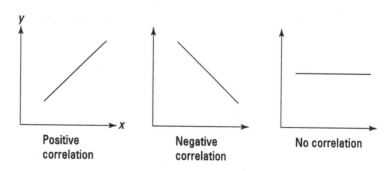

Ask your child to consider the following data for age and IQ:

Age	IQ
5	100
10	120
15	115
20	98
30	135
40	101

Using the following chart, ask your child to plot the data:

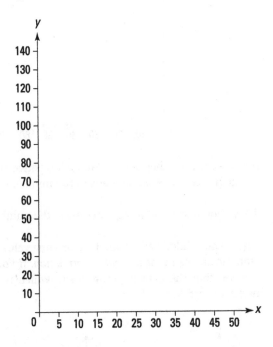

They should get:

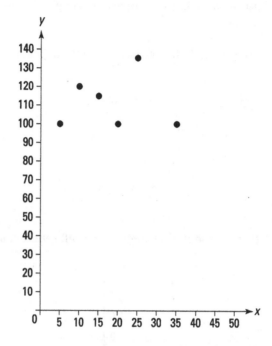

Ask your child if they see any type of relationship between age and IQ. They should say no — the data does not go up (positive correlation) or down (negative correlation). So, in this case, there is no correlation between age and IQ.

Sample Size Matters

In the previous section, your child learned to determine if a correlation exists between data values. For simplicity, those examples used a small amount of data, meaning a small sample size or population.

REMEMBER

Using a small sample size can be misleading. When more data is considered, the result may change.

In this section, you discuss the importance of sample size with your child.

Say to your child: "You have learned to use a scatter plot to determine if a correlation exists between data values. To make the previous charts easier to understand, each one used a small set of data points, which statisticians call the *sample size*."

Explain to them: "When you use data to consider correlation, the more data you use — meaning the larger the sample size — the more accurate your results will be."

Ask your child to consider the following data that compares hours playing video games to age:

Age	Video Game Hours
5	1
10	2
15	3
20	3.5

If you graph the data points, you will get the following:

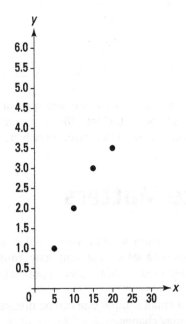

Say to your child: "Looking at the graph, you would tend to say there exists a positive correlation between age and the hours spent playing video games. However, it's important to note that your sample size is small: 4."

Explain to your child: "If you were to consider a larger sample size, the chart would change as follows."

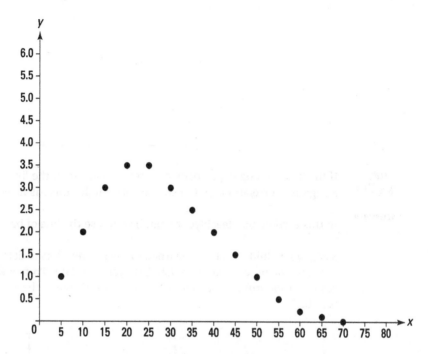

Say to them: "In this case, the larger sample size indicates there is a negative correlation between age and hours spent playing video games."

Tell your child: "As you work with data, you should always consider the sample size."

Using a Line of Best Fit

In the previous section, your child used scatter plots to determine if a correlation existed between data values. To make such correlations easier to identify, statisticians often draw a line, called the *line of best fit*, in the data.

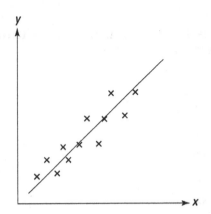

If the line is going up, a positive correlation exists. If the line goes down, there is a negative correlation. And, if the line is straight, there is no correlation.

In this section, you teach your child how to use the line of best fit.

Say to your child: "You have learned to use scatter plots to determine if a positive, negative, or no correlation exists between the data. To make such correlations easier to identify, statisticians draw a line through the data called a line of best fit."

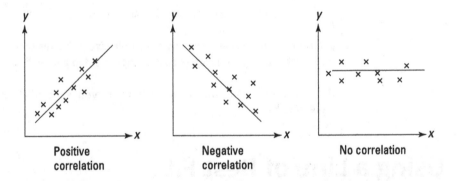

| Positive correlation | Negative correlation | No correlation |

Say to your child: "Using the line of best fit, you can easily see if a correlation exists. If the line goes up, a positive correlation exists. If the line goes down, a negative correlation exists. And if the line is straight or goes up and down, no correlation exists."

Working Through Word Problems

Word problems — they never get old, do they?

Present the following word problem to your child:

> Billy's rugby club asked the members to sell cookies to help the club raise money. The club kept track of cookie sales by member age. Use the following chart to plot the data:

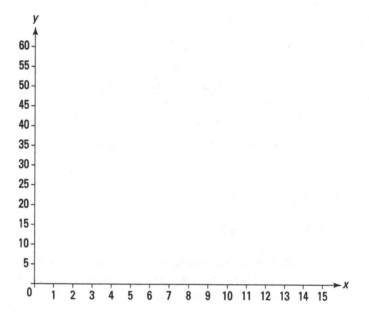

> Draw a line of best fit through the data and determine if there is a correlation between cookie sales and age.

THE INFAMOUS TRAIN-STATION PROBLEM

If you ask people about word problems, the infamous train-station problem often comes to mind:

> Billy boards a train heading west and Mary boards a train heading east. Billy and Mary are 300 miles apart. Billy's train travels at 45 miles per hour and Mary's train travels at 55 miles per hour. How long will it take Billy and Mary's trains to cross paths?

(continued)

(continued)

1. **Determine the average speed of the trains.**

$$\frac{55+45=50}{2} \text{ miles per hour}$$

2. **Tell your child: "If the trains are going the same speed, they will cross paths after 150 miles — the halfway point."**

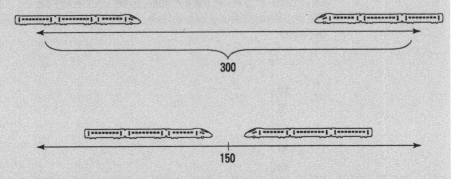

3. **Determine the amount of time it will take the trains to travel 150 miles.**

$$\text{Average speed} = \frac{55+45}{2} + \frac{100}{2} = 50 \text{ miles per hour}$$

$$50\overline{)150} \text{ (3)}$$
3 hours

4. **Remind your child that Mary's train is faster than Billy's train. So, after three hours, the trains will pass each other, but Mary's train will have gone farther. Draw the following pictures:**

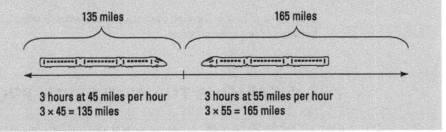

3 hours at 45 miles per hour   3 hours at 55 miles per hour
$3 \times 45 = 135$ miles $3 \times 55 = 165$ miles

Present the following word problem to your child:

Mary's class takes a survey to determine how many hours of TV the students watch at night.

Student	Hours of TV at night
Mary	3
Bill	5
Juan	3
Shenise	2
Gabriel	5
Sally	2
Jane	1

What is the average number of hours that the students watched TV?

1. **Have your child add up the hours of TV watched (21), and divide it by the number of students (7):**

   Average $= 21 \div 7 = 3$ hours

2. **Using the following chart, graph the frequency of each hour watched (for example, 2 students watch 5 hours of TV).**

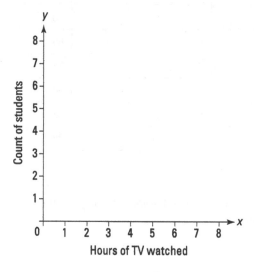

3. They should get:

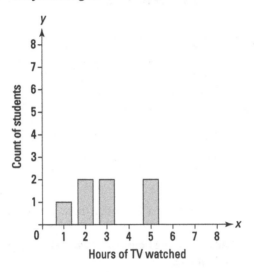

The students took a math test on Friday. The following data shows each student's score and the number of hours they watched TV:

Student	Hours of TV at night	Score
Mary	3	70
Bill	5	65
Juan	3	72
Shenise	2	80
Gabriel	5	68
Sally	2	78
Jane	1	85

Ask your child to use the following chart to graph the test scores and hours of TV watched:

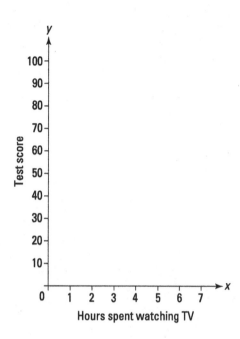

They should get:

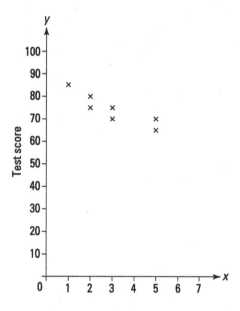

Ask your child if there is a correlation between the amount of TV watched and how the students scored on the test.

They should identify a negative correlation.

5

The Part of Tens

Chapter **29**

Ten Types of Mathematics

Now that your child has successfully mastered math through eighth grade, you may be wondering, "What's next?" To address that question, this chapter examines 10 different types of math your child may encounter in high school or college.

Pre-Algebra

Pre-algebra is often called "middle-school math" because it targets seventh-, eighth-, and ninth-graders. Many of the concepts presented throughout this book are considered pre-algebra, including (but not limited to):

» Integers, decimals, and fractions

» Negative numbers

» Roots and powers

» Scientific notation

» Probabilities

» Statistics

Students normally don't get to choose whether they take pre-algebra. Everyone gets to take pre-algebra. The good news is that I call it the "math for life" because it teaches the math you use every day.

Algebra

Algebra is a branch of mathematics that uses letters such as x and y within expressions to represent variables. The following are examples of algebraic expressions:

$$x = 5 + 3$$
$$y = 3x + 1$$

This book provides your child with a start in algebra. Again, all high-school students have to take algebra. Often, a student's success in algebra influences other math courses they take in the future.

TIP

If your child begins to struggle with algebra, find a tutor who can help. Students who do well in algebra tend to take more advanced math.

Many careers such as computing programming, engineering, accounting and finance, and medicine use algebra on a daily basis.

Business Math

Business math consists of the math a business uses to manage its operations, forecast sales, manage inventory, and more. Common operations in business math include basic math, algebra, and probability and statistics. For advanced business operations, business math can include calculus.

Students who plan to work in the future will work for a business. Understanding business math will give them a leg up.

Geometry

Geometry is one of the oldest types of mathematics. Geometry deals with spaces (shapes), distances, angles, areas, volumes, and proofs. Several of the topics presented in Chapters 9, 20, and 27 are concepts from geometry.

A wide range of jobs use geometry, such as engineers, architects, carpenters, plumbers, and designers (both interior and fashion).

Trigonometry

Trigonometry is a branch of mathematics that examines the relationships between the lengths of the sides of a triangle and the triangle's angles. Trigonometry focuses on the sine, cosine, and tangent of angles.

Careers that require trigonometry knowledge include engineering, computer science, mathematics, architects, crime scene investigators, and astronauts.

Pre-Calculus

Pre-calculus is not necessarily a type of math, but rather a collection of math topics that students should know before taking calculus, such as:

>> Trigonometry

>> Geometry

>> Exponentials

>> Logarithms

>> Rational functions

Students who plan to take calculus will take pre-calculus to gain a solid math foundation needed for more advanced math. Many college majors, such as engineering and computer science, will require pre-calculus if not calculus.

Calculus

Calculus is a branch of mathematics that studies change. There are two primary branches of calculus:

>> *Differential calculus* studies rates of change.

>> *Integral calculus* studies the area under a graph curve.

Calculus has a wide range of applications, from finance to space travel. In finance, calculus is used to determine how fast different companies are growing (or shrinking). In space, it's used to determine the rate in which two objects are coming together or the velocity needed to achieve a specific order.

Other careers that use calculus include engineers, mathematicians, physicists, computer programmers, and even animators.

Probability

Probability is a branch of mathematics that studies how likely an event is to occur. For example, using a six-sided die, the probability of rolling the number 1 (or any other number on the die) is 1/6. Probability is used for a wide range of purposes, from the stock market to gambling.

Many businesses use probabilities to determine things like the potential success of new products or the success of a business merger or acquisition. Your child will use probabilities in the following careers: medicine, marketing, engineering, biology, firefighting, crime solving, and more.

Statistics

Statistics is a branch of mathematics that studies data to interpret, analyze, and visualize it. Simple statistics include the average value, variance, median value, and mode. Scientists make extensive use of statistics to measure the results of their experiments.

Careers that use statistics include computer science, data science, business analysis, engineering, medicine, biology, and actuarial science.

Linear Algebra

Linear algebra is a subset of algebra that deals with *linear expressions* (expressions that graph to a line). The simplest form of a linear expression is $y = mx + b$. Businesses and data scientists make extensive use of linear algebra to chart data and to predict future trends (predictive analysis).

Other careers that use linear algebra include engineering, computer science, medicine, biology, business and financial analysis, accounting, loan and insurance underwriting, mathematics, and physics.

Chapter **30**

Ten Things to Consider before High School

This chapter examines ten things you and your high schooler or soon-to-be high schooler should consider.

Study College Options

Different colleges have different degree programs, job-placement statistics, and costs. Out-of-state schools are normally more expensive than in-state colleges, and four-year schools are more expensive than community colleges. Take time to study your child's college opportunities. Request catalogs from each school (most catalogs are also available online). Understand their entrance requirements, costs, and scholarship offerings. Visit, if possible.

Consider Trade Schools

College is not for everyone. As an alternative, you should examine different trade schools, such as engine mechanics, plumbing, HVAC repair, and more. Again, you want to know the entrance requirements and costs. Many trade schools have job-placement programs that can lead to successful careers.

Take the SAT Early and Often

Many colleges use SAT scores as part of their entrance process. Students often find that they increase their score by taking the exam multiple times. They should take the exam early and often and use their results to identify potential courses they can take to improve their score.

Study a Foreign Language

Many colleges require a foreign language for entrance and also for graduation. Your child can often meet these requirements by taking a foreign language in high school.

Participate in Meaningful Extracurricular Activities

High school offers a wide range of extracurricular activities, many of which look good on college and job applications. From sports, to band, to clubs, your child's choice of extracurricular activities can say a lot about how they spend their free time. Colleges are obviously interested in students with good grades. When two students have similar grades, the colleges will lean toward the student with the more well-rounded background.

Take Advantage of College Courses

Many colleges offer courses to high-school students at little or no cost. Your child should take advantage of such courses. Not only will the courses look good on their college application, but the credits may also let your child skip courses at a higher price when they become a college student. Check out available courses from your local colleges. Talk with the guidance counselor at your child's school to find out about available opportunities.

Remember that Grades Matter

After your child starts high school, every grade becomes part of their transcript and can impact college admissions, scholarship opportunities, and more. With good grades come honor rolls and national societies, each of which looks good on college and job applications.

Learn about Potential Scholarships

College is expensive. Fortunately, many potential scholarships are available to a wide range of students. Take time to research early. Many good books and websites are available that can help. For example, an ROTC scholarship normally covers all four years of college expenses. Know your scholarship opportunities and take advantage of them.

Study Career Paths Early

Students have many career options, and most high-school students have no idea what they want to do in life. Help your child research a wide range of career options. Their interests may lead them to take different courses in high school or select a different college. Have them talk with their school's guidance counselor, who may be able to provide materials on different careers or point them to online surveys that will question them about their interests.

Take Advantage of Tutors

Depending on the classes your child chooses, they may find some classes challenging. In such cases, seek out a tutor — perhaps an older student at the school or a college student looking to earn extra money.

Index

Symbols and Numbers

< (less than) symbol, 146
> (greater than) symbol, 146
π (pi), 138, 317
0 (zero)
  exponent of, 198–199
  multiplicative property of, 264
1 (one)
  exponent of, 198
  not a prime number, 122

A

absolute value, 153–153, 289
acute angles, 271–272
acute triangles, 278–279
addition, 13–29
  checking their own work, 81
  counting cards, 17
  decimals, 103–105
  exponents, 329–330
  flashcards, 14–15, 17–19
  fractions
    like fractions, 85–87
    unlike fractions, 91–93
  negative numbers, 147–148
  number lines, 22–26
    three-digit addition problems, 25–26
    two-digit addition problems, 23–25
    worksheets, 24, 26
  properties of
    associative property, 265
    commutative property, 264–265

identity property, 263
  regrouping (carrying and borrowing), 18–22
    three-digit addition problems, 19–22
    two-digit addition problems, 18–21
    using boxes, 18–19
    without boxes, 20
    worksheets, 19–22
  timed addition tests, 15–17
  word problems, 27–29
    simple, 27–28
    three-digit addition problems, 29
    two-digit addition problems, 28–29
    worksheets, 15–17, 19–22, 24, 26
additive inverse, 151–152
algebra, 207–223
  defined, 386
  inequalities, 218–222
    interpreting, 221–222
    representing visually, 219–221
    solving, 218
    using, 218
    worksheets, 220–222
  one-variable expressions, 208–214
    complex expressions, 209–211
    expressions that include multiples of x, 211–212
    performing multiple operations to solve for x, 212–214
    simple expressions, 208–209
    worksheets, 208–212, 214

  one-variable expressions with exponents, 215–217
    complex expressions, 216–217
    simple expressions, 215–216
    worksheets, 216–217
  simplifying expressions, 214–215
  two-variable expressions, 305–313
  unknowns, 207–208
  word problems, 222–223
  worksheets, 208–212, 214, 216–217, 220–222
angles, 269–281
  defined, 270
  fun angle-related facts, 273
  names of, 274–275
  protractors, 270
  relationships, 275–278
  solving for unknown angles, 276–278
  triangle types, 278–281
  types of, 271–273
  worksheets, 277–278
area, 134–137, 354–356
  of circles, 139, 141, 205
  of parallelograms, 354–355
  of rectangles, 135
  of squares, 135–136
  of trapezoids, 357
  of triangles, 136
  word problems, 140–141, 362–364
  worksheets, 137, 139–, 356
associative property
  of addition, 265
  of multiplication, 267

multiplication, 49–66
box multiplication, 57–62
process for, 57–59
three-digit multiplication problems, 62–64
two-digit multiplication problems, 60–62
worksheets, 62
checking their own work, 81–82
counting cards, 51, 53
decimals, 106–109
exponents, 202–203, 327–328
flashcards, 50–51, 53, 55, 59–60, 120
fractions, 88–89
negative numbers, 149–150
old-school math, 53–57
three-digit multiplication problems, 55–57
two-digit multiplication problems, 54–55
worksheets, 55, 57
properties of
associative property, 267
commutative property, 266
distributive property, 267–268
identity property, 263–264
timed multiplication tests, 51–52
word problems, 64–66
three-digit multiplication problems, 66
two-digit multiplication problems, 65–66
worksheets, 51–52, 55, 57, 62
multiplicative property of zero, 264

N

National Governors Association, 9
negative exponents, 200–201, 326–327

negative numbers, 143–159
absolute value, 152–153
addition, 147–148
additive inverse, 151–152
charting negative values, 153–158
comparing with positive numbers, 146–147
division, 150–151
explaining, 144–145
multiplication, 149–150
number lines, 144–148, 151–152, 157
origin of concept, 146
subtraction
of negative numbers, 148–149
that results in negative numbers, 144–145
worksheets, 148–149
word problems, 158–159
worksheets, 147–151
new math, 8–9
being open to, 9
encouragement when child makes mistakes, 11
helping children to learn, 10–11
investing in child's future, 10
opinions of, 9
origin of, 9
routine for learning, 11
Nine Chapters on the Mathematical Art, The, 146
nonlinear equations vs., 335–340
number lines
addition, 22–26
three-digit addition problems, 25–26
two-digit addition problems, 23–25
inequalities, 219–222
negative numbers, 144–148, 151–152, 157

subtraction, 41–45
three-digit subtraction problems, 43–45
two-digit subtraction problems, 41–43

O

obtuse angles, 271–272
obtuse triangles, 278–279
old-school math (old math), defined, 9
one (1)
exponent of, 198
not a prime number, 122
one-variable expressions (solving for *x*), 208–214
complex expressions, 209–211
with exponents, 215–217
complex expressions, 216–217
simple expressions, 215–216
expressions that include multiples of *x*, 211–212
moving to two-variable expressions, 306–307
performing multiple operations to solve for *x*, 212–214
simple expressions, 208–209
worksheets, 208–212, 214
online resources
Cheat Sheet (companion to book), 3
worksheets (companion to book), 3
order of operations, 262–263

P

parallelograms
area of, 354–355
defined, 352
perimeter of, 352
PEMDAS acronym, 262–263

subtraction, 31–48
  checking their own work, 81
  counting cards, 33
  decimals, 105–106
  exponents, 329–330
  flashcards, 32–33, 35–36
  fractions
    like fractions, 87–88
    unlike fractions, 93–94
  negative numbers
    subtraction of negative numbers, 148–149
    subtraction that results in negative numbers, 144–145
  number lines, 41–45
    three-digit subtraction problems, 43–45
    two-digit subtraction problems, 41–43
    worksheets, 42–45
  regrouping (carrying and borrowing), 34–40
    three-digit subtraction problems, 37–39
    two-digit subtraction problems, 36–37
    using boxes, 36–37
    without boxes, 39–40
    worksheets, 37–40
  timed subtraction tests, 33–34
  why not commutative, 266
  word problems, 45–48
    simple, 45–46
    three-digit subtraction problems, 48
    two-digit subtraction problems, 47
  worksheets, 33–34, 37–40, 42–45

T

timed tests
  addition, 15–17
  division, 69–70
  multiplication, 51–52
  subtraction, 33–34
tipping, 228–230
trade schools, 389
train-station problem, 377–378
trapezoids
  area of, 355
  defined, 352
  perimeter of, 352
triangles
  area of, 136
  perimeter of, 133
  Pythagorean Theorem, 357–359
  types of, 278–281
  worksheets, 281
trigonometry, 387
tutors, 391
two-variable expressions (solving for x and y)
  charting expressions, 321–324
  process for, 305–313
  worksheets, 307, 313

U

unit rates, 249–254
  calculating, 249–252
  commonly used, 252
  defined, 249
  word problems, 253–254
unknowns, 207–208, 308
unlike fractions
  addition, 91–93
  defined, 91
  subtraction, 93–94

V

variance
  calculating, 285–287
  defined, 286
volume, 359–362
  of cones, 361
  converting cubic feet to gallons, 362
  of cubes, 359–360
  of cylinders, 360–361
  of spheres, 360
  word problems, 364
  worksheets, 361–362

W

Wallis, John, 146
word problems, 222–223
  addition, 27–29
    simple, 27–28
    three-digit addition problems, 29
    two-digit addition problems, 28–29
  decimals, 117–118
  division, 78–80
  fractions, 97–99
  metric system, 245–246
  multiplication, 64–66
    three-digit multiplication problems, 66
    two-digit multiplication problems, 65–66
  negative numbers, 158–159
  percentages, 235–237
  probabilities, 299–301
  ratios, 193–194
  shapes
    area of, 362–364
    perimeter of, 362–364
    volume of, 364
  statistics, 179–184, 287–289, 377–381
  subtraction, 45–48
    simple, 45–46
    three-digit subtraction problems, 48

About the Author

Dr. Kris Jamsa is the author of 119 books, mostly on computers and programming, but also several on pre-K through fifth-grade learning, including the Wiley book Teaching Your Kids New Math, K–5 For Dummies. Jamsa has two PhDs — one in computer science and another in education — and five master's degrees (business administration, project management, computer science, information security, and education). He has an undergraduate degree in computer science from the United States Air Force Academy.

Jamsa is the founder of Head of the Class, an online learning portal for pre-K through fifth-grade learners. His research interests include all things computer, as well as ways in which people can use technology to improve learning.

He and his wife Debbie live on a ranch in Prescott, Arizona, with their dogs, cats, and horses.

Dedication

To Debbie: My love for you is infinite: Love = ∞

Author's Acknowledgments

Making a book on math easy to understand takes a great team of editors, illustrators, and designers — a hard process. I want to thank the Wiley team for doing a great job. I also want to thank Kathleen Kelley, this book's technical editor, for her focus on Common Core standards. In addition, I can't compliment Marylouise Wiack, this book's copy editor, enough for her great contributions. Marylouise did much more than fix my grammar. She often asked great questions and provided real-world examples that made this a much better book. Finally, Chrissy Guthrie, this book's development editor and editorial project manager, kept me on track, kept the process fun, and improved every aspect. Chrissy's contributions appear on every page of the book. I sincerely appreciate all of their efforts and contributions.

Publisher's Acknowledgments

Senior Acquisitions Editor: Jennifer Yee

Project Manager and Development Editor: Christina Guthrie

Copy Editor: Marylouise Wiack

Technical Editor: Kathleen Kelley

Production Editor: Mohammed Zafar Ali

Cover Image: © damircudic/Getty Images